MAKING
SIMPLE MODEL
STEAM ENGINES

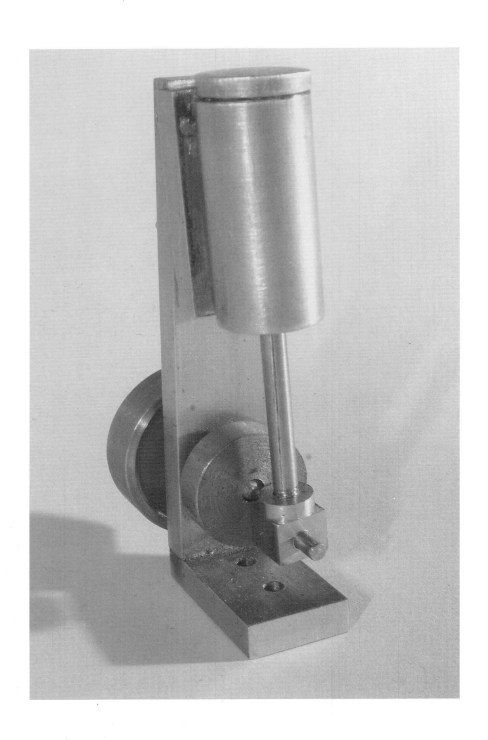

MAKING SIMPLE MODEL STEAM ENGINES

STAN BRAY

THE CROWOOD PRESS

First published in 2005 by
The Crowood Press Ltd
Ramsbury, Marlborough
Wiltshire SN8 2HR

www.crowood.com

This impression 2010

British Library Cataloguing-in-Publication Data
A catalogue record for this book is available from the British Library.

ISBN 978 1 86126 773 3

Disclaimer
Safety is of the utmost importance in every aspect of metalworking.
When using tools, always follow closely the manufacturer's
recommended procedures. However, the author and publisher
cannot accept responsibility for any accident or injury caused by
following the advice given in this book.

Designed and typeset by Focus Publishing, Sevenoaks, Kent

Printed and bound in India by Replika Press Pvt. Ltd.

Contents

PART ONE
INTRODUCTION

Getting Started

Steam engines led the industrial revolution; their introduction into industry changed the way goods are produced, and therefore subsequently the way in which human beings live. Some will say this is a pity, as life now has a far quicker pace than it did before the industrial revolution, but others feel that the advantages of mechanization far outweigh the many disadvantages that it brought with it. Of course generally the steam engine as a means of power has long been superseded by the use of oil-based fuels and even nuclear power. Even so, much of industry is still powered by steam except that these days it tends to be used in turbines rather than the old-fashioned reciprocating engine.

Basically machines operated by steam are simple: it is a case of boiling water, collecting the steam that is generated in the process, and putting it in a closed cylinder so that it will operate a piston. The trick in being successful is the manner in which the steam is allowed in and out of the cylinder, and there are many methods of doing this, some of which are extremely simple, others that can be quite complicated, the latter engines being developed in the endless search for greater efficiency by the engine designers. This book will only deal with comparatively simple mechanisms; nevertheless, anyone making one of them will build a working engine.

So how does a steam engine work? In this modern world we have electrically operated kettles that switch themselves off as soon as the water boils, but not so very long ago the kettle was boiled on a stove, and when it boiled it was taken off to make the tea, or for whatever purpose. If you forgot the kettle had been put on and left the room, you would come back and find the room absolutely full of vapour, usually referred to as steam, although in fact it is actually the steam condensing. No matter; what it actually is is the result of steam having been generated, but the remarkable thing is that it has completely filled the whole room from a quantity of about a half litre of water. Thus there has obviously been a vast temporary expansion in the quantity, although when it condenses it will return to more or less the original amount.

If instead of leaving the room, the person who put the kettle on remains and perhaps dozes off, they would have been awakened by the sound of the lid of the kettle bouncing up and down. What is happening is that some of the steam cannot escape quickly enough and is pushing the lid off as it expands; obviously then the expansion of the steam is generating some force. It is generally believed that it was this that caused James Watt to develop his interest in steam power, although in fact this is unlikely. Steam had already been in use for many years, and his contribution was to develop and improve the engines that were already in use or being built.

The first steam engines that were put to practical use were built by Thomas Newcomen, and were designed purely as pumps for use in the mines in Cornwall. At that time the full potential of steam had not been realized, and the engines worked on the principle of applying steam to move the piston and then applying cold water to create a vacuum and thus to draw it back from whence it came. The engines were purely reciprocating the connecting rod moving in and out in a straight line.

The fact that the cylinder was cooled each time meant that the engines were very wasteful of steam, and a considerable improvement came about when instead of cooling the cylinder, the used steam was allowed to exhaust. This meant that two taps – or 'cocks' as they are always referred to in steam parlance – were required, one to allow steam in, the other to allow the exhaust out. These were operated manually, a task usually given to boys, which must have been boring in the extreme, particularly as the working day would have been twelve or even fourteen hours. It is not recorded how many times the lad doing this operation might have fallen asleep or lost concentration, and in doing so would have brought the whole of the pumping operation to a standstill; the mine owner would have been less than pleased.

We can only speculate how long it took, or who it was that first decided that if they connected the two taps to the piston rod of the engine with rope or wire, the taps would open and close by themselves. No doubt in doing so the person responsible gave themselves a nice easy time until what they had done was discovered, when they would probably have been sent elsewhere to work. Whoever it was, they unknowingly invented the first-ever valve gear, and should be hailed as a genius instead of being forgotten, as well as having possibly received punishment.

Even though a primitive valve gear had been discovered, the engines still only worked in a reciprocating movement, and they were still inefficient, and it wasn't until the crank was invented – something attributed to James Watt – that it became possible to put the steam engine to other uses. The crank enabled the reciprocating motion to be converted to a rotary one, and by adding a heavy flywheel it was then possible for an engine to run continually; moreover the rotary motion could also be used to drive pulleys that operated machinery via belts. From being just a machine for pumping, the steam engine was now a means of powering all sorts of machinery, and it did not take industrialists long to realize the fact.

A feature of the old mills and engineering works was the numerous overhead shafts driven from an engine, each carrying a number of pulleys that in turn were connected to a machine of one sort or another. It was all very noisy, and if the engine stopped for any reason the whole of the mill or factory would grind to a halt; nevertheless it was the beginning of high volume production and the industrial revolution.

Having obtained rotary motion, it was a comparatively short step for it to be adapted to driving a vehicle, and almost immediately attempts were made to fit steam engines to road vehicles, to railway engines and to boats, all of which is well documented. Although very powerful, the steam engine has many disadvantages, and so it was inevitable that other forms of prime mover should eventually displace it for most industrial purposes.

The reciprocating engine has all but disappeared, and where steam is in use, these have been replaced by turbines that are steam driven but work on an entirely different principle. The turbine finds many uses, not least the generation of electricity, and that electricity is usually now used to provide power to operate the more modern versions of the pumps and other machines that were driven by the reciprocating steam engine.

Children like to play with toys, particularly those that represent modern artefacts. Come to that, a considerable number of adults also like to have such toys – except in their case they are referred to as 'models'. With the industrial revolution in full swing, it was logical that some enterprising manufacturer would quickly make a steam-driven toy for the benefit of children, and in fact it was not just one manufacturer, but a considerable number that produced a range of them. From the early to the mid-nineteen hundreds they were what we now refer to as 'best sellers'. Most were stationary engines of the simplest form, but there were models as well as toys, and some of the engines were used to drive a range of other models, possibly the most popular of these being steam locomotives, which in those days were the main form of power for public transport. A number of these models were produced to a high standard and it was also possible to buy parts with which to assemble one's own model, although complete kits of parts were very rare indeed.

As the years went by, other forms of power became common, and with the development of small electric motors the interest in steam engines declined. In all probability this was because turning on a switch was far easier than the effort involved to run the steam engine. In addition, the materials from which toys were made was rapidly changing, so whilst at the turn of the century all mechanical toys were made from sheet metal folded to shape and painted accordingly, the advent of die-casting with low melting-point alloys changed all that, and not only enabled the toys to be produced much more cheaply but also to be more realistic looking. Many of the manufacturers of steam toys and models went out of business or

converted to the new methods, with just a handful remaining to make steam-driven toys or models; at least two of these are still enjoying success with them, so the interest has never completely disappeared.

With the exception of a few more sophisticated models, the steam engines were usually of the type known as oscillating engines, and it is this type of engine with which this book mainly deals. The oscillator in full size was used in industry and in particular for marine work, and yet the principle is so simple that a model is very easy to make, and when well made can produce a considerable amount of power; it is even possible to make one with hand tools, such as a file or two and a hand drill. There are examples of such models in this book, although it is mainly about simple models that can be made with the use of a lathe. All the designs can be built using one of the very small machines, such as a Unimat or similar design.

The models vary in complexity, starting with something very simple and developing into more complicated designs. One does not need to be a highly skilled engineer to build them, and in doing so the knowledge of metalworking practices will be gradually improved. The materials needed are simple and basic and are generally easily obtainable; model engineering suppliers will stock everything that is required. In fact it is not necessary to obtain the materials from the latter, as most metal stockists will be able to supply all that is required, and they frequently have scrap boxes that will provide the required metal, and for only a nominal sum. For a few pounds, therefore, a model can be made that would possibly cost nearly a hundred to buy, and in addition there is the fun and satisfaction of knowing it has been built by oneself.

MEASUREMENTS

About fifty or so years before this book was published, it was agreed that Great Britain should conform to the European Continent, as well as many other countries, by adopting the metric system of measurements. This was in principle a very noble idea but unfortunately had not been properly thought out, as virtually every machine being used by industry was designed for working with the British imperial measurement. When using a machine such as a lathe or milling machine the operator works to measurements contained in graduations on dials. If the dial is designed to give a reading in imperial measurement, it is not practical to use it for working to metric figures, without a major rebuild of the whole machine. Just changing the dial for a metric one will not work either, as the graduations are directly related to the pitch of the threads used to move the various slides. For example, where one turn of a handle might move a slide on the lathe $\frac{1}{10}$", the nearest practical figure we can get to that in metric is 2.55mm. Being no longer a round number addition of other metric units, or the multiplication thereof, sees us finish with some very awkward decimals.

The result is that Britain continued to use the imperial system, and as a large part of our trade was with the United States of America, a country that has no intention of changing from their form of imperial measurements to the metric system, many, if not most, companies carried on as they always had. The manufacture of articles in imperial measurements demands the use of metal bars, sheet and so on to that same system, and therefore materials also continued to be produced to the old standard. However, some companies that were trading with countries other than the USA soon found it necessary to convert to machines that had metric facilities, and to use metal stock to that standard; others continued using the imperial standard, and made use of material produced to the old standard. Even today as this book is being written both standards are still used, the metric form only now becoming widely used.

The result has been quite a mess, and this has had its effect on model makers as well. All old drawings were imperial, but after a while some new designs started to appear in metric form, and the model engineer was piggy in the middle. We therefore have the situation where some model makers have imperial equipment, and others metric, and each needs drawings in the form that he or she uses. Conversion using a calculator is not a practical answer, because whichever way the conversion is done, the results end up with decimal figures impossible to work to. This means that each design has to be tailored to the measurements used by the individual, resulting in two separate models of quite different sizes. This situation remains today, with some metal stockists only selling to the old standard, some all metric, and others a bit of each.

It is a situation that creates great problems in a book of this nature, because if designs are produced to just the metric format, then half the model makers or prospective model makers will find it of absolutely no use whatever; and vice versa. It has therefore been necessary to try and offer something for everyone by using both systems, and allowing the reader to decide which he or she wishes to work to.

In many cases it is possible for a single drawing to contain both measurements, and readers will find a number of instances of this. The metric figure will be shown as a number in brackets, except where diameters and radii are concerned, where the number will be followed by the letters mm. Imperial figures will be followed by the sign ", thus hopefully making all the measurements absolutely clear. In many cases it is just not practical to include both sets of measurements on a single drawings or part of a drawing, and so two drawings of the same model have been produced in different formats. Where this happens the metric figures are shown without brackets. It is not usually possible to interchange, therefore a model started in metric format should be completed that way, the exception being in the case of a skilled operator who will have the experience to know what parts are interchangeable.

The above applies to making the models in general, but there are exceptions, and certain individual items may possibly be inter-changeable between the two systems. As well as measurements of components there is also the question of threads, and once again in general the advice is to stick to one system if possible. However, taps and dies used for thread making are expensive items, and it might suit some people to use, say, an imperial thread when they are building a metric model, because they happen to own suitable threading tools. There is nothing wrong with this, and suitable tools and materials when already available can be used; for example, there is no reason why a length of brass bar to imperial measurements cannot be machined or filed to a metric size. In the case of threads, however, beware of the fact that the use of a different format may have an adverse effect on other components, and so it will be necessary to think some way ahead when converting in this fashion.

As far as our American friends are concerned, it is unfortunately not practical to produce yet another set of drawings to suit their engineering standards, but the imperial drawings should prove to be quite satisfactory. The only difference will again be in the thread arrangements, the imperial ones here being quoted as British Association (BA), while the USA use the Unified System. Fortunately, between the 'coarse' and 'fine' specification it is possible to get nearly identical to the BA sizes, and conversion will prove to be no great hardship.

SAFETY

It is expected that anyone engaged in the operation of machinery will conduct themselves in a safe manner, and wear appropriate clothing in the interests of safety. In particular this applies to the use of safety glasses, gloves and footwear.

The models described make use of methylated spirits, a highly inflammable liquid, and care must be taken when using it, in order to avoid injury. It can be particularly dangerous when spilled and allowed to ignite, as it burns with a flame that is nearly colourless. If in any doubt it would be wise to use special tablets that can be purchased at any model shop and some hardware stores; these should be put on a suitable small tray, and this placed where the burner should be.

The tablets will operate all the models shown, though they give less heat than the methylated spirit burners and so models will run a little more slowly. They also unfortunately quickly deposit a layer of soot and oil mixture on the surface being heated, which acts as insulation and therefore needs to be cleaned regularly. Young children should not be allowed to operate the models except under the supervision of an adult, and in the case of operation by children alone, the tablets should definitely be used, rather than methylated spirit.

As long as care is taken, the use of methylated spirits is not dangerous, and thousands of models similar to those described in this book have been made and are operated on a regular basis. Methylated spirits is also commonly used by youth organizations as a fuel for cooking.

Readers of this book who construct the models described herein, or indeed anyone involved in making or operating similar models, must ensure that the practices they employ are safe. Neither the author nor the publisher accepts any responsibility for damage or injury caused to, or by, any person making or operating any of the devices described herein. It is the responsibility of the person concerned to ensure they are working or operating in a safe manner and environment.

Drawing Conventions

For the benefit of any reader who is not conversant with the various conventions used in engineering drawing practice, featured below are those that are likely to be found within these pages.

Outline of object or component of that object

Centre line A line that is an indication only of the relative position of two points. It is not constructional but is used as an aid to construction.

Hidden detail Indicates the position of a component that is behind something shown as an outline. Hidden detail may go to several depths or be left out, depending on how valuable the information is it is believed to supply.

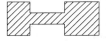

Sectioned drawing
Shows component as it would be expected to appear if sliced across at a certain point. Particularly used where too much hidden detail can make things illegible. Angled lines can go in either direction and frequently do so where two mating items are sectioned. This may also be shown by using different line spacings.

Dimensioning lines
Show distances between two points

abc

Angular dimension

26.64°

Alignment dimension

0.32"

Radius dimension

R 0.59"

1 General Construction Notes

TOOLS

Some of the projects in this book are small engines that can be made using only hand tools, and this idea can easily be expanded to build more complicated ones, still using just hand tools. So let us give some thought to the tools that we need both for those, and for the more advanced models that follow.

Most people reading this book will already have access to such things as a screwdriver, pliers, the odd file or two, and hopefully some marking-out tools as well. In addition a small blowlamp is desirable, but before dashing out to buy one it would be advisable for the less experienced reader to first read the paragraph about them in the chapter on boilers (*see* p. 102).

A drill of one sort or another is also required, together with suitable drill bits; most households nowadays have a power drill of one sort or another, and while these save a lot of hard work, they can also very easily ruin the work, so extreme care must be taken when using one. The problem with many of them is that they have only one or two rotational speeds, which are too fast for our purpose; and while they will certainly drill the required holes, as a rule they will not drill them very accurately. This is because it is almost impossible to hold them still while drilling, and the excessive speed turns this movement into holes that are not exactly round, and are frequently of a larger diameter at one end than the other.

Some drills have an electronic control that

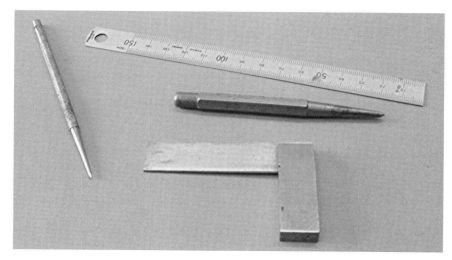

Selection of some marking out tools that will be useful when making the models; they include a scriber, rule, centre punch and engineers' square.

will allow them to be used at lower speeds; nearly all re-chargeable ones have this facility, and these tools are of a great deal more use for our purpose.

For the small amount of drilling required to make one of the simpler engines described herein, an old-fashioned hand drill has many advantages and much to recommend it. Of course, the ideal tool for drilling is a properly designed drilling machine.

Marking out tools are also desirable; a scriber and rule are a must, although the former can be improvised by filing a fine point on a nail, as in general only soft material will need to be marked. A centre punch of sorts can also be improvised from a nail if needs be, although in the case of both the scriber and centre punch it is much nicer to have the proper tools. For some models, although a pair of dividers and a small square might appear to be an added luxury, they are more or less a necessity, and will certainly allow work to be carried out with greater accuracy than when improvised tools are used.

For anyone wishing to make the more complicated models, a lathe is essential: in general, one of the miniature or compact types of machine will suit, but it is necessary to ensure it has a four-jaw chuck with independent adjustment to each jaw. A tailstock drilling chuck in which to hold drills and taps is also something that will prove its worth in a very short space of time, and readers with lathes who do not have this facility are advised to get one as soon as possible

As far as the actual drills, taps and dies needed are concerned, this will depend entirely upon the model or models being constructed; it is not worth buying complete sets, but it is best to get just those required. There must be millions of taps and dies sitting around in the workshops of model makers, which have never been used and never will be, because there is just no call for certain sizes. There is also the question of whether the reader wishes to work to imperial or metric measurements; both are catered for, but it is not advisable to mix them,

and far better to stick to one standard. It is always best to buy top quality drills, taps and dies, since cheap ones frequently do not cut accurately and in addition break easily. All of these three are best purchased at a tool supplier, and not a DIY store.

To use the taps and dies it will be necessary to have a tap wrench and die holder. While tap wrenches can be bought, it is sometimes better to make them for oneself; very little effort is required, and the end result is a tool that suits the individual rather than one that has to be adapted to. In particular, small diameter taps require small tap wrenches; frequently those sold are far too large and cumbersome, and therefore so unwieldy that they have a tendency to break the taps.

A small vice is essential, as is a suitable work area. The work will not involve creating a great deal of mess, but using the kitchen table might not be ideal as brass filings give a rather odd taste to a meal. Although an experienced metal worker might complete a model in a day or so, many people will need to spread the work over a number of hours. There are numerous small parts to be assembled and it will be essential to have a container of some sort in which to house them.

'D' Bits

A number of references are made in the text to the use of 'D' bits: these are a tool much used in model engineering but rarely available commercially, and it is unlikely that they will be available from more than one or two suppliers, and certainly not from tool stockists. The reason for this is that the tool is very simple to make, and in doing so can be tailored to individual requirements. The 'D' bit is used to flatten the end of a drilled hole that will have been left angled as a result of the shape of the drill. Different-size 'D' bits will be required for several of the projects.

To make them one needs a short length of silver steel of the required diameter: at one end a short length is filed flat to half the diameter of the steel, and the front of that is tapered by

'D' bit, showing how the round stock is reduced to half diameter.

Another view of the 'D' bit in which the cutting angle is clearly visible.

about 10 degrees. The steel is then heated until it is the colour of a boiled carrot and immediately quenched in water, which makes it very hard. The problem now is that it is too hard and therefore brittle, and so it needs to be tempered in order that while remaining hard, the brittleness will disappear. It now has to be cleaned until it returns to its original brightness, for which emery paper is used, then it is re-heated once more. However, on this occasion the heat is applied away from the end that has been worked on, and the steel will be seen to slowly change colour. These colours will move along to the prepared end, and when the tip of that becomes pale blue it is again quenched in water. If possible, a rub over the cutting edges with a small oilstone will help to sharpen it.

SOFT SOLDERING

Soft soldering will be necessary, and it is not all the type of work that a soldering iron will cope with, so a small blowlamp will be essential; it must, however, have sufficient capacity to supply the required amount of heat, and the modern type sold in DIY stores where the gas container screws into the nozzle is fine. If possible it should be of the type that gives a pointed flame, rather than a wide, spread one. The use of lead-free solder is strongly urged: not only is it less of a health hazard, but it also melts at a higher temperature than most of the leaded ones, something that is very useful when making models to be driven by steam.

SILVER SOLDERING

Silver soldering, or silver brazing as it is sometimes known, is very similar to soft soldering, but much higher temperatures are involved. Whereas most soft solders melt at a temperature range of between 180 and 300°C, depending on the particular type and brand, the melting point of the range of silver solders starts at around 550, rising to over 900°C. The first consequence of this is that the use of a soldering iron is ruled out, and a blowlamp must be used.

The general principles otherwise are the same as with the use of soft solder. Both surfaces of the work to be joined must be well cleaned, and flux applied liberally. A dip in acid will remove most of the oxidization that will have occurred during the time the metal has been stored, and it can then be cleaned with either a fine file or an abrasive cloth or paper. Only a light rubbing is needed to remove any unwanted dirt, and all dust created by the cleaning must be completely removed. Many people prefer to clean the metal with a stainless-steel scraper as no unwanted residue is left, unlike abrasive papers, which leave fine abrasive grains behind. A suitable scraper can be made from an old kitchen knife or a piece of stainless-steel rod. Because it is only to be used around the area to be joined, the rod need not be very large: generally a piece of $\frac{3}{16}$" or 5mm rod will do quite well. A flat blade must be filed on the end that is to be used for cleaning the metal. A very good alternative to the scraper is a stainless-steel wire brush, but under no circumstances should an ordinary steel brush be used, as these are likely to leave a residue that will prevent the joint being clean enough to solder.

The purpose of any flux is to avoid the metal oxidizing, and it is essential to use a flux that will do this at the melting temperature of the solder. Most fluxes for silver soldering are in powder form, and although it is quite acceptable to sprinkle the dry powder along the proposed joint, it is usual to mix the flux with a little water so that a thick paste is formed, and spread that along instead. The water aids the flux to penetrate the joint, and this penetration can be improved still further with the addition of a little washing-up liquid. If the temperature rises too quickly the flux will sometimes boil away from the joint, so it is essential to heat the job slowly. The flame of the blowlamp should never be played directly on to the joint, but always at the side in such a position that the conductivity of the metal will cause the area at the joint to rise to the correct temperature. Flux also has a nasty habit of running where it is not wanted as it boils; this is something that is difficult to stop, although drawing a line with a lead pencil around the joint area and about ⅛" or 3mm from it will generally contain the solder within that area. This still depends to some extent on the temperature being raised, as sometimes overheating will cause the flux to actually run beyond the pencil control line.

As the temperature rises the flux will be seen to change viscosity, first taking on a soft texture, and then returning to a solid state, at the same time changing colour. It is usually at the latter point that the silver solder can be applied directly to the joint area where the flame is not touching. Trying to apply it at the point where the flame is will only result in a blob of solder falling off the end of the stick, and remaining thus.

Once silver solder has been melted, as indeed the blob of solder has, the temperature required to re-melt it is considerably higher than that needed in the first place. This means that the work will need a prolonged heating otherwise the blob of solder will simply remain where it is, stuck to the parent metal. We are then left with two choices, either to file it off, or to allow it to remain in situ where it will look most unsightly and do nothing to increase the strength of the joint. Such blobs do not occur if the solder is melted on the parent metal close to, but not at the point of, flame contact.

Silver solder has two stages of melting, technically known as the solidus and the liquidus, although the names need not bother us. The first means it is at a plastic sort of state, the second, as the name implies, that it is fully liquid. It is the second stage that is the effective one, though for anyone silver soldering for the first time it is all too easy to assume melting has taken place when the first stage is seen to have occurred. It is essential that it is allowed to become fully liquid, otherwise the joint will almost certainly fail.

When large areas are silver soldered, as is the case when boiler making, it is possible for air pockets to appear in the silver solder. Usually this is due to the flame being moved too quickly along the joint area, resulting in a tiny place where the solder does not run to, because the temperature is not quite high enough; or it can be that too high a temperature has boiled the flux right off. Such places can be spotted as soon as they occur and should be dealt with prior to the solder solidifying; as has already been pointed out, allowing that to happen will require a considerably higher temperature in order to affect a repair. These small areas can soon be sorted out using a length of thin stainless-steel rod, about ⅛" or 3mm diameter, that has been sharpened to a fine point. Simply rake it into the liquid solder: this will clean off any oxidization, and the solder will follow and flow into the joint.

After silver soldering the flux must be washed off, not only to improve the appearance of the work, but also to prevent possible corrosion. Most fluxes will wash off in cold water, but occasionally they will need to be chipped away. When cold water does not do the job, it may be necessary to use hot water with a little washing-up liquid. Liquids such as white spirit, paraffin and turpentine are unlikely to have any effect. Sometimes difficult residues can be removed with methylated spirit. After removing the flux, the work can be immersed in a mild acid; citric acid is particularly suitable as it is non-toxic and can be safely disposed of after use. Small parts can even be cleaned in neat vinegar if one wishes to save the chore of mixing an acid bath.

MATERIALS

All the materials required to make the models in this book are easily obtained from the suppliers of model engineering equipment. In addition some can be obtained from model shops, and it is also possible to purchase material from metal stockists. It works out much cheaper to buy from a metal stockist, although very often they are not willing to supply the small quantities required. Occasionally even a long length of material might be cheaper than a shorter length of the same material bought from another source; the best idea is for several like-minded enthusiasts to get together and share the spoils.

Most of the material used will be brass, with a small quantity of stainless steel used mostly for connecting rods. Copper tubing is generally used for the pipes to connect the engines to the power supplies, and copper is used for boilers. Boiler fittings and suchlike are best made from bronze, for reasons that are explained elsewhere.

THE MODELS

Steam toys are always sold as complete models, in the form of an engine, boiler and burner; some include other additional models for the engine to drive, and others are finished as transport models. A different pattern is followed in this book, with separate descriptions of the engines, boilers and burners; all of which are of a more substantial form of construction than is generally found in steam toys. They are not described as individual models, the choice of which engine to join to which boiler being left to the builder.

There is a good reason for this approach, in that constructors may possibly not wish to have a vertical engine with a vertical boiler, but may instead wish to fit a horizontal boiler to the engine. All that has to be done is to build whichever engine and whichever boiler is wanted and link them together; it is as simple as that.

In addition, boilers found on steam toys vary very little between manufacturers, and hardly at all between individual models by a particular company. Following this pattern would mean that the same boiler is described over and over again with a number of engines, so with a couple of exceptions any of the boilers can be fitted to any of the engines. The boilers themselves vary in complexity, giving a newcomer to this sort of construction the opportunity to start with something very simple, and to then progress to a more difficult form of construction when they are ready.

Some people may like making the engines, but are not too keen on building boilers, and that is fine: build one boiler and connect it to each engine as and when it is wanted, to drive that particular model. It is even possible to build a single boiler and to connect it to a manifold that in turn is connected to several engines, allowing them all to be run at once, or individually as one wishes. Such a manifold can, if one wishes, be connected to an air compressor and the engines operated on compressed air, there being no need at all to build a boiler if one does not wish to do so. A little compressor such as those used to power an airbrush will do the job, or even an old refrigerator motor can be pressed into service and will provide sufficient power to operate one or two models at a time.

The final finish is also left to the whim of the individual; most people appear to be quite happy to connect boiler and engine together, with the boiler on a stand, and to operate it like that. Others want to fashion some form of engine house, in some instances carrying out quite detailed modelling on the building and accessories. Whatever the individual requirement the models can easily be adapted to suit.

There is also, of course, the person who will build just for the pleasure of making things, operating them being a secondary consideration, if it is considered at all. As they are all constructed of brass, copper and bronze, they will polish up well for display if that is what one wants, and they can look very presentable indeed.

COMMON DENOMINATORS

Certain items are common to all models; in particular these include boiler connections and fitting, which are dealt with as appropriate in Chapter 12 (boiler construction). The method of cylinder and frame construction for oscillating engines is also very similar, and therefore is not repeated in every chapter. The other common item is the flywheel, and although it is something every engine needs, it is also a matter of personal choice as to the form it takes; the subject of flywheel construction is therefore dealt with in this chapter.

FLYWHEELS

The purpose of the flywheel is to add momentum to the engine and keep it going. A piston reciprocating in a cylinder stops at the end of each stroke, and although engines are designed to minimize the effect, without some additional help they will work in something of a stop-start fashion, and possibly will stop running altogether. One of the ideas that James Watt introduced to improve the running of steam engines was the introduction of a flywheel. The rotation of the flywheel carries the engine beyond the position where it has momentarily lost its inertia, giving it a smooth operation and possibly even making up for faults in design that cause the loss of momentum. In order for the wheel to rotate, the piston has to be connected to a crank,

which is an offset from the shaft on which the wheel runs; this converts the reciprocating motion to a rotary one, which is generally far more useful.

It follows therefore that making a flywheel should not be a hit-and-miss affair, with any old circular piece of metal doing the job. The wheel must be of sufficient weight and diameter for its purpose, while at the same time not being so heavy that the power of the engine is absorbed by the action of driving the flywheel. It is also essential that it runs perfectly true, for while an out-of-true wheel will still give some momentum to the engine, it will also have the effect of causing some unevenness in the running, as the weight will tend to give sideways as well as rotary movement.

The engines described herein will work very well with a fairly small diameter flywheel, made of solid brass or steel. Aluminium, of course, is not suitable, as it does not have sufficient weight to provide the required momentum. In general a large diameter wheel is not necessary, so construction will be quite simply a case of machining a hole accurately in the centre of a piece of round bar; the best way to ensure that it is true and accurate is to machine a little from the outside diameter, and drill the hole at the same setting.

For the sake of appearances it is a good idea to machine small recesses in the face of the wheel, and of course some means of securing it to the crankshaft will be necessary. Machining the bar to leave a central boss that protrudes slightly beyond the width of the rim can do this; the boss is then drilled and tapped to accept a small grub screw that secures the wheel to the shaft. If this idea is thought a little unsightly, the hole can be made a push fit to the shaft, or the wheel can even be secured to it using one of the various adhesives that are available. It is not a good idea to use a high-strength retaining compound, as it may well be necessary to remove the flywheel at some time, and although the bond of such adhesives can be broken by the application of heat, this may well result in damage to the model.

In full-size practice, the wheel will be keyed to the shaft, and this is a method usually adopted on larger models of stationary engines. It is much more complicated than using a grub screw, and in addition the keys required for the tiny models described here would be so small that it is not a method to be recommended. Nevertheless if anyone wishes to make the effort it is possible to do so.

The appearance of some engines may well be improved with a large flywheel; a main feature of engines used for pumping, and of those used in the mills, was the enormous wheels. However, it must be remembered that a large wheel rotating at a high speed will become unstable, and as a general rule the rotational speed of the engines was very slow indeed. This applies equally to a model, and therefore care must be taken not to run an engine with a large, heavy flywheel too fast.

If a larger-than-usual, solid flywheel is to be fitted, the area between the central boss and rim must be machined to a very thin section in order to gain an overall reduction in weight, and to ensure that most of it is at the circumference of the wheel where it will have most effect. The alternative is to make a wheel with spokes, which can either be done by filing the spokes from solid material, or by inserting them through the rim and into the boss. Doing the latter involves making a simple jig, involving nothing more than cutting a couple of circular recesses in a block of wood; the way to do this will become apparent when the accompanying photographs are examined.

As well as solid and spoked flywheels, there were examples in full-size practice with round or elongated round sections cut from the web; this was done to reduce weight. Anyone wanting to use this sort of design for a flywheel in model form will certainly find it is quicker than making a spoked one, and at the same time achieves the object of reducing overall weight, while concentrating a large proportion of what weight there is at the periphery.

A typical, small steel flywheel of the type used by most of these engines; note the recess in the face, which has been made to improve the appearance.

Making wheels in this way will, of course, only apply to the larger ones, but any engine looks better with a large flywheel, and providing that increasing the size does not make it too heavy, there is no reason why the size should not be increased on any of the engines, and suitable arrangements made to obtain the necessary clearance by adjusting the base.

For one model where a large flywheel was deemed important for the sake of appearance, a wooden flywheel with a cardboard rim has been constructed. This has been fixed to a larger than usual central boss, the latter acting to give the momentum. It is not an ideal arrangement, as the weight is concentrated in the centre of the wheel rather than at the outside; nevertheless the idea does work and looks effective, which is all that matters.

Finally, yet another alternative is to purchase and machine a suitable casting, a variety of which is available from suppliers of model engineering materials. Remember, however, that a casting will almost certainly be heavier than a fabricated flywheel, and this should be taken into account.

To sum it all up, a flywheel should not only maintain the momentum evenly, but must also run true; if this is not so, uneven running can result. The way to ensure absolute accuracy is to carry out as much machining as practical at one setting.

PART TWO
THE ENGINES

2 Hero's Engine

As the first known example of a device driven by steam was designed by Hero of Alexandria, it is fitting that we start this section with a model based on his idea.

Who really invented the steam engine is shrouded in mystery, details of many devices having been lost in time. A common fallacy is that it was James Watt who got the idea from seeing how steam lifted the lid of a kettle. He may well have conceived ideas from that, but he most certainly did not invent the steam engine; to him we owe many of the advances that were made in its development, but attempts to harness the power of steam were made hundreds of years before that. For example, it is known that Giovanni Branca of Lorento in Italy built an engine powered by steam in 1629: sketches of it depict a jet of steam from a statue being blown on to a wheel with fins, though whether it actually took that shape we cannot say for certain; but he definitely did build a machine working on that principle. And there were a number of others at around the same period who also attempted to harness the power of steam. Most of these primitive machines consisted of a boiler that looked rather like an embellished kettle, the steam that came from the spout being used to drive a fan-type arrangement, which we would now know as a primitive turbine.

Giovanni Branca was not by any means the first person to try the idea, and written records exist of a man called Hero of Alexandria in or around the year AD50, who produced a machine consisting of a sphere with two spouts through which steam passed, and because of the shape of those spouts, the steam drove the sphere in a rotary motion. It was probably little more than a novelty item and not capable of doing any amount of work, and because the records are so sparse, nobody is quite sure what the machine really looked like. A number of artists' impressions have been produced over the years showing several different ideas on how the machine might have looked. These vary, from a simple sphere supported on two pivots and holding water that was heated by a fire underneath it, which created the steam and drove the machine, to other pictures that show the engine having a much more complex shape, where one side has features that are a cross between an animal and a human being, the steam being ejected from the mouth of this figure, causing it to rotate. Yet another far more complex version depicts the sphere supported over a double-headed animal holding a fire. The fire in turn heats a boiler, and steam passes through the arms that act as pivots and thus into the sphere and in turn out through a pair of spouts. The latter arrangement would be the most efficient, except that keeping the pivots steam-tight where they join the sphere would be very difficult, and one would think near impossible, with the facilities then available.

We cannot therefore truly reproduce the Hero engine, but the two shown in this chapter work on the same principle, and both are a good exercise in simple boiler-making techniques as well as providing interesting working models. Readers will of course have noticed that neither model in this case is a sphere, but both are tubes sealed at each end. Making a sphere that will hold steam is far from easy and not the type of exercise in keeping with the contents of this book. However, both work quite well and can be powered by an ordinary candle, so without burners to make, they are real 'quickies', and each can be made in a couple of hours or so.

It will be seen that the difference is mainly in the position of the pivots: number one has the pivots in a horizontal position (see drawing/photo), whereas in the case of number two they are vertical. There is some difference in the running qualities, in as much as number one tends to have a slightly less smooth motion than number two. This is caused by the small imbalance created by the filler cap, which does not exactly counterbalance the jet. As it happens, this is a useful feature as it means that while the water in the cylinder is warming up there is no possibility of the jet falling to the bottom, and thus allowing an escape of steam too early. While number two has a better balance it is marginally more difficult to make, but nevertheless is still not hard. Both engines can possibly be made from scrap material.

ABOVE: A copy of an early sketch showing the principle of Hero's Engine and how it was believed to have been.

RIGHT: Engine Number One, which is shown here, differs from Number Two in as much as one is horizontal, the other vertical.

ENGINE NUMBER ONE

Although the engine in the photograph was made from a short piece of scrap brass tubing, a short length of larger-diameter central-heating pipe can be used for the body, barrel or tank, whatever one wishes to call it; pieces of the same tubing can be opened out and flattened to form the ends.

Before soldering the ends in place, drill the two holes on the circumference, otherwise the expansion of the air during heating will prevent a good joint being made. Hard or silver soldering is recommended, but if facilities for that are not available, soft solder will do, as there will not be sufficient heat to melt it. Don't try and shape the ends perfectly before soldering them to the tube, but solder roughly shaped pieces in place, and trim round with a file to make a neat job; then the barrel can be put in the lathe and the necessary holes drilled for the end bearings. If no lathe is available, carefully find the centre of the end by measuring, and make an indentation using a

centre punch or something similar; repeat this on the other ends. These are the bearing dimples.

A short length of copper tube is soldered in one of the two holes on the circumference of the barrel, and is bent round to roughly form a right angle. The end is pinched either in the vice or with pliers so that only a very tiny opening is left. If the point of a domestic pin is put in the end when it is pinched together, that should leave a hole small enough to create a powerful jet of steam.

The other hole has to have an arrangement whereby water can be put in the barrel; this should be made with a matching plug to seal it. Once again, for anyone without the use of a lathe there is no need to despair: simply solder on a suitably sized brass nut, and use a brass screw for sealing.

The two supporting arms are made with lengths of steel strip riveted to short lengths of angle; there is a single tapped hole in each piece of bar, and the holes must be in line with each other when the engine is assembled. Points are

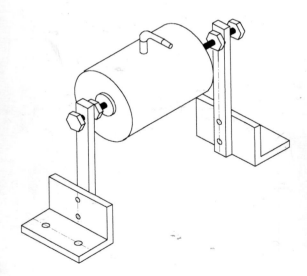

General view of the Hero Engine Number One. It should be mounted on a suitable base such as a wooden block.

The bearing arrangement for engine Number One. Note it is essential that adjustment of the bearings is such that the engine will rotate freely when the merest touch is applied.

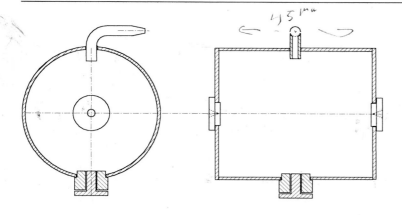

Hero-1-Barrel - copper tube as used by plumbers 32mm dia. 1-3/4" (45)
long. End fittings-brass approximately 3/16" (4) dia. Exhaust tube from
1" (25mm) length of 1/8" (3) dia. copper tube. Filler cap and plug
2BA (4mm dia.) or similar-brass nut and screw.

Hero Engine Number One.

ABOVE: Barrel.

RIGHT: Support brackets: two required from solid steel or brass.

*BELOW: Pivot screws: two off from stainless or mild steel. Thread
to match pillars.*

BELOW RIGHT: Fixing brackets: two off from any angle material.

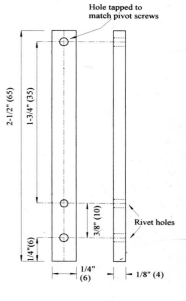

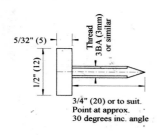

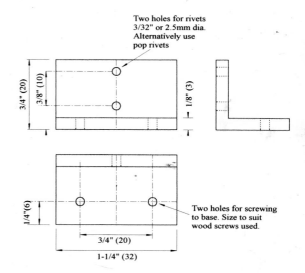

made on ordinary screws to act as bearings, and a lock nut is desirable to stop them unwinding when the engine is in motion.

The bearing screws should be adjusted so that the weight of the filler plug draws it to the bottom, no matter what position it is put into, and a light touch will spin the barrel on its axis. The whole assembly is screwed to a suitable block of wood.

About a teaspoon or two of water is all that is required to cause the engine to rotate. When it stops, do not try and remove it immediately as copper retains the heat and it will be far too hot to handle.

ENGINE NUMBER TWO

As usual, measurements are shown on the drawings, but within reason they can be adjusted so that any handy material can be used. The body of the engine in the photograph was made from a length of old central-heating tube. The ends were made from the same material, a couple of pieces being cut lengthways and after softening, flattened on the bench. They are silver soldered to the tube and the surplus trimmed afterwards. Note that they should not be flanged, because in order for the model to operate successfully weight has to be reduced as much as possible.

As before, while silver soldering is desirable, you can assemble the parts with soft solder, safe in the knowledge that the amount of heat will not melt it. Start by soldering in place the end that will be the top of the engine, then trim it and file it round the circumference to make a neat edge. Drill a suitable hole centrally in the end plate to accept the bearing; this is best done in a lathe if one is available, otherwise careful marking out and centre punching will do. The bearing can be made from brass or bronze.

The top bearing also acts as the filler plug,

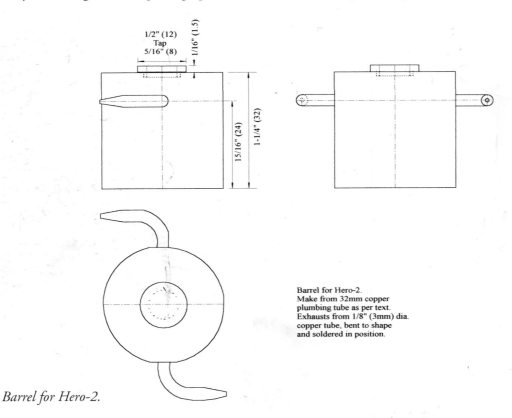

Barrel for Hero-2.
Make from 32mm copper plumbing tube as per text. Exhausts from 1/8" (3mm) dia. copper tube, bent to shape and soldered in position.

Barrel for Hero-2.

The barrel shell of engine Number Two, showing the holes drilled ready to accept the jet arms.

The top of the barrel has been completed, and the position of the bearing/filler fitting can be clearly seen.

Top cap: one off brass or bronze.

1/2" (12) dia.

3/8" (10mm) dia. thread
5/16" (8mm)

1/16" (1.5)

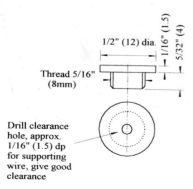

1/2" (12) dia.

1/16" (1.5)

5/32" (4)

Thread 5/16" (8mm)

Drill clearance hole, approx. 1/16" (1.5) dp for supporting wire, give good clearance

Top bearing – also used for filling.

The bearing is in place, as are the arms that drive the engine.

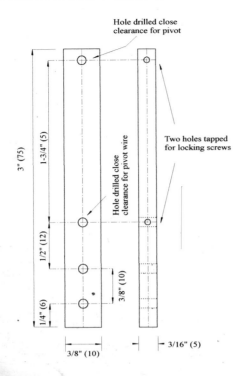

Support pillar: one off mild steel or brass.

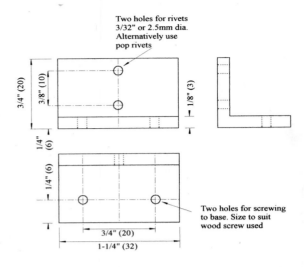

Fixing bracket: one off from any angle material.

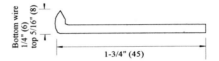

Supporting wires for barrel top. Point should be longer than the one at the bottom.

and a screw will be required with which to seal it. In the centre of the screw a hole is drilled part way through; this assists with putting the barrel on and off the stand. It can be soldered into the end at this stage, but take care not to melt the solder holding the end to the tube when doing so.

Soften the piece of copper to be used as a base, and with a hammer give it a good old belting; this will spread it and thus thin it, and the thinner it is, the better the heat conduction properties. Instead of using a piece of the tubing for the base it would be advantageous to use copper shim, as sold in art shops for decorative work, as that is a very thin-section material.

Finally solder the base in position, and trim

it to size and shape. When that is done, make a tiny indentation in it; this can be done with a centre punch, or you can put the whole assembly in the lathe and bring the tailstock centre to it with just sufficient pressure to dent it – don't go too hard, as there is a danger of puncturing the copper, which at this stage is very soft. Check that it runs true, as the indentation will be used to locate the needle bearing when the engine is finally assembled. It is obvious that a great deal of care must be taken to ensure that the indent really is true, otherwise the engine will develop an alarming wobble in operation.

Making the frame should be straightforward enough, and once again measurements can be adapted to suit individual circumstances.

It will be seen that it consists of a short length of angle (either iron or alloy will do), and this is riveted to a length of mild steel bar. There are two holes in the bar, just a little longer than the length of the cylinder, and there are tapped cross-holes in one side to meet these. Two lengths of wire have points filed on them and are then bent at right angles. They are eased into the holes so that the pointed ends of the wire just touch the bearing dimples. The top angled section is slightly longer than the bottom one and the clearance hole in the top cap is for the excess to go in. Once in the hole, the point should touch the point at the end in

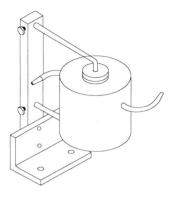

Vertical version of Hero's Engine.

The completed engine, showing the method of mounting it on the vertical frame.

27

order to get a very free-running bearing. (In the model in the photograph the wire was actually a cut-up coat hanger.) Two screws secure the wire in the required position, and prevent it twisting during operations. If for any reason the two bearing dimples are not exactly in line with each other, adjustments can be made to the wire to get things lined up a little better.

Before final assembly it will be necessary to tweak the exhaust pipes to shape, using a pin to obtain the necessary jet when the ends are closed. Not only does this give a jet, but because of the small orifice it also prevents too much steam escaping before the engine is ready to start, which is sufficient to allow the heat inside to build up during operations. Should the diameter prove insufficient it is an easy enough task to just push the point of a pin into it to open it out, at a later date. When assembled, a light push on one side of the barrel should cause it to spin for several rotations.

To operate the engine, the barrel is removed from the stand by undoing the screw holding the top arm in position, and twisting it sideways; it can be returned by using the reverse process. The bearing cap is removed, and about a teaspoon or two of water is poured into the barrel through the hole; note that it is no use trying to fill the thing up because then it just will not work, as there would be no room for the steam to expand. It would also need a considerable increase in the heat source to boil that amount of water.

Put the barrel back in the bearings, and rotate the engine to ensure it is running as true and easily as possible; then light a candle underneath, and when steam starts to appear from the exhaust, give the engine a tweak and away it will go. Performance can be improved by making a small methylated spirit burner to use instead of the candle, and that will also allow a little more water to be put in and therefore the engine will run for longer.

3 Simple Sam

The techniques required to build this engine are described in detail, and the methods required to build most of the other oscillating engines are exactly or very nearly the same. Rather than repeat the instructions in every chapter, readers can refer back for any information on construction when building other engines of this type; obviously where there are major differences these will be dealt with as and when necessary. In addition, a number of detail differences in construction are also explained, in order that readers can in some ways customize this model if they so wish. Many of those variations will also apply to other engines in the series, and that side of the construction will be left to individual choice.

This is a basic oscillating engine, a single-acting, single-cylinder type that has been popular with model makers for over a hundred years. At one time many steam plants using oscillating engines were sold commercially; these were aimed not only at the market for model enthusiasts, but also at children. A few such models are still available on the toy market, although with modern technology most mechanical toys are now powered by an electric motor. Even so, steam still appears to have a fascination, and children invariably enjoy watching them.

The basic oscillator like this is easily made, and doing so is a suitable project for a youngster, offering a good, yet simple introduction to engineering and science. Young people like to get things running quickly, something that is possible with this model because the type of construction is very

forgiving, and a couple of minor mistakes in measurements will usually make little or no difference – the engine will still work. To build it requires the use of a small lathe, as well as general hand tools, and a small blowlamp is also necessary to build the boiler, a simple basic vertical type.

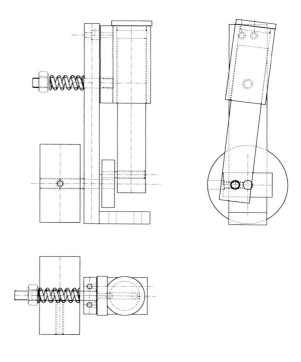

A general arrangement of Simple Sam showing completed assembly.

THE CYLINDER

The model has a built-up cylinder assembly, for which a length of tubing can be used – although tubing frequently does not have a very smooth bore, so it will be necessary to lap it in order that the bore can be made reasonably presentable. Making a lap is not complicated, at least not a simple one like this: it just consists of a length of round steel bar with a saw-cut down the middle; the end of the cut is sealed with some epoxy resin.

A short length of emery cloth is then pushed into the slot, and with the lap held in the tailstock chuck, and the length of tubing in the headstock chuck, the lathe is slowly rotated so that the emery cloth smoothes the bore. Do not be tempted to run the lathe fast: nothing will be gained, and a great deal of unwanted heat will be generated. Also, do not allow the emery to keep protruding from the ends of the tube; doing so will result in bell mouthing, which is the rounding of the ends of the tube. A small amount of unevenness, although not desirable, can be tolerated, so there is no necessity to lap the tubing for any great length of time. If anyone is working without a lathe, the tubing can be held in a bench vice and the lap slowly

rotated with a hand drill. There is no objection to using a power drill providing it will operate at a slow speed.

Some people will prefer to drill a length of solid rod, and if so, bronze can be used as an alternative to brass. The material can be drilled in either a three-jaw chuck, or a collet, whichever suits the builder best. There are some people who even prefer to use the four-jaw chuck for such a purpose, working on the principle that if set up with care, it is far more accurate than the three-jaw version, virtually all of which seem to have some eccentricity. Drilling will give a perfectly suitable bore, but a more accurate and better finished one would be made by using a reamer to finish off.

The Cylinder Cover

In full-size practice, cylinder covers were always bolted in place so they could be easily removed if required. However, it would be extremely difficult to attempt to bolt the cover on this little engine, and so the design calls for it to be soldered in place. It is simply a piece of brass rod of the same outside diameter as the tube, with a step machined to allow it to fit into the tubing; it should not be such a tight fit that force is needed for it to be pushed home. At the

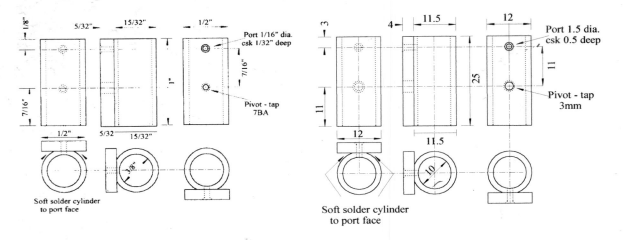

Cylinder block and port face: fabricate from brass.
LEFT: *imperial measurements;* RIGHT: *metric measurements.*

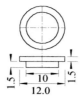

Cylinder cover: one off – brass. LEFT: *imperial measurements;* RIGHT: *metric measurements.*

A piece of tube is used for the cylinder, but before it is used the bore has to be smoothed to remove ridges that are formed when the tubing is made. The cap is machined from brass rod, and will be soldered in place.

same time, it must not be too sloppy, either. The drawings call for the step to be 1.5mm in length but there will be no harm in making it slightly shorter than this if one so wishes.

Those making the cylinder from solid metal may wish to make a blind bore so that the cover is part of the cylinder; this involves making a blind bore. Making a blind bore is not all that difficult but in order to get the end a 'D' bit will be required. Readers making the model without a lathe can solder a piece of brass sheet across the top of the cylinder, and then file it to match the periphery.

The Cylinder Port Face

The cylinder port face is a short length of flat brass with two holes in it, one for the pivot, the other to act as the steam port. It is also advisable to make a shallow groove lengthways to locate the cylinder in, when joining the two together. Failing that, it will be necessary to make a flat on the cylinder itself and use that as the mating surface.

Start by scribing a line centrally along the length of brass, then *gently* centre-punch the position where the pivot hole is to go. Set a pair of dividers to ¾" or 20mm and, with one end in the centre-punch mark that has just been made, scribe an ark to cross the line marked along the centre, and centre-punch the point of intersection. Drill a 1/16" or 1.5mm hole at that point, ensuring that it is square, and remove the burr on the side that is to be soldered to the tubing. The other side needs to be very slightly countersunk: the drawings call for this to be 0.5mm deep, though slightly less will do no

The port face is marked out, the port drilled and the pivot hole drilled and tapped. To get a good mating surface for the cylinder a radiused groove is made.

harm; if the countersink is too deep the engine will not operate at full power.

The pivot hole has to be drilled and tapped, and it is essential that both hole and thread are square to the face of the material; if there is any doubt about one's ability to get the tap square,

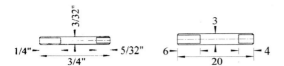

Pivot: mild steel, thread ends 7BA or similar.
LEFT: *imperial measurements;* RIGHT: *metric measurements.*

which is something that is far from easy, make a small guide and use that. The guide is only a short length of mild steel bar with the end machined square and a clearance hole for the tap. By holding the square end on the work and passing the tap through it, a square thread is guaranteed.

Pivot Pin

The pivot pin is a short length of stainless steel threaded at each end; the shorter of the two threads fits into the port face, the other will be used to adjust the tension of the assembly at a later stage. The pin should, if possible, be threaded in the lathe: most readers will no doubt have a tailstock die-holding device, but if not, use an ordinary stock type of holder and support it against the tailstock barrel in order to ensure the thread remains square.

Joining Port Face and Cylinder

The port face and cylinder are soft soldered together, and the pivot should be fitted before doing so in order to prevent the tapped hole from being filled with solder. However, it is desirable, if possible, not to solder the pivot pin to the face, because if this happens, cleaning the port face becomes extremely difficult. This can be achieved by rubbing an ordinary soft lead pencil into the thread and across the end, since soft solder will not adhere to a surface covered with the graphite from the pencil.

Coat the surfaces to be joined liberally with flux, and then assemble them and hold them together with soft iron wire. Usually a couple of turns of wire in one position is suitable, though some people might prefer to use two. At one time soft iron wire could be bought in small coils at any oil shop or ironmongers, but nowadays such premises are rapidly disappearing and their place has been taken by DIY stores, which do not stock this type of article. However, ironmongers do still exist, and it will generally be found that they will have some in stock. It is a commodity used by florists, and it might be worth trying to get some from that source. Failing that, the core of the common binding material made of plastic and used for holding small pieces together

The pivot consists of a length of stainless steel threaded at each end, before the port face is soldered to the cylinder, the pivot should be screwed in place. To prevent the hole from becoming filled with solder, prior to the soldering operation the thread that fits the port face is covered in lead pencil so that it will not accept solder; then the pivot can be removed if necessary.

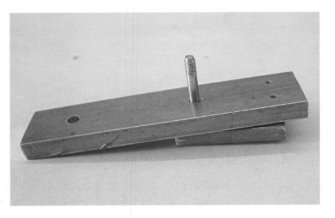

This photograph shows how the pivot pin will protrude through the steam block when the engine is completed.

ABOVE: The port face has been liberally fluxed, it now remains to screw in the pivot and then assemble to the cylinder.

RIGHT: The components to be soldered together must be bound together with soft iron wire. One source of such wire is the flexible plastic binding used to tie up bags; the outer coating of plastic is easily stripped off.

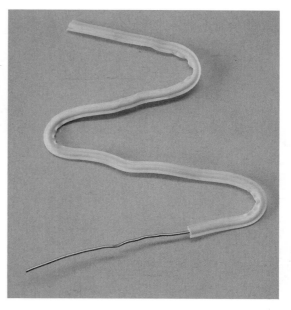

when they are purchased can provide a source, and iron wire can be stripped from that. Whatever happens, do not be tempted to use copper or brass, as it will join to the work, whereas the iron will not.

The cylinder assembly can be recessed between the ports and pivot; this helps to maintain a good contact between the two mating surfaces of port face and port block, as frequently metal strip that has the appearance of being flat is not. The recess takes care of possible undulations as just the two short sections are mating, rather than the full length of the port face. However, it may be as well to run the engine for a while without the recess, and if there is no blow of steam, to let well alone. After all, there is no point in making extra work that is not necessary.

Fitting the Cylinder Cap

It makes sense to fit the cylinder cap at the same time as the port face and cylinder are joined, so put flux all round the lip of the cap and push it home. We are now ready for the soldering operation, which is best carried out with the cylinder cap at the bottom. The work should be

stood on a piece of heat-proof material; at one time we would have used asbestos, but now it will be necessary to find an alternative. Do not be tempted to use a metal block, as that will absorb all the heat that is needed for the soldering operation. It is possible to buy a small quantity of special material from model engineering suppliers or companies that specialize in soldering equipment; mostly the material is a silicone sheet, which is white and not unlike asbestos to look at, though there is another material that looks just like a piece of hardboard and is sold under the trade name of Skamolex. Before commencing to heat the work, put a tiny piece of solder in the cylinder so that it lies as close as possible to the periphery.

With a small blowlamp, heat the assembly at the end that is away from the cap, and wait until the piece of solder that was dropped inside has melted and run round the edge. The job should now be hot enough to apply solder to the edges of the port face, and it will run down this and form a small fillet along the length. It should be possible to do this without applying any more heat; however, should this not be so

33

The port face is secured to the cylinder with the wire, and the rim of the cap fluxed; the assembly can then be heated with a small blowlamp and soldered together.

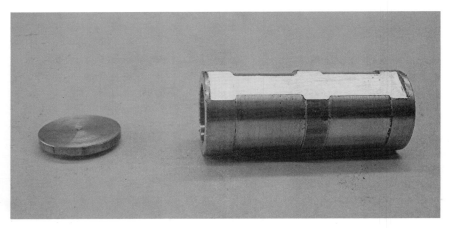

Anyone not confident about getting an accurately radiused groove in the port face can instead file or machine a flat on the cylinder and obtain a good fit that way. Doing so means that the crank will need to be thinned in order to line it accurately to the cylinder bore.

and it is necessary to re-introduce the flame, make sure that not too much heat is applied, as that can cause a bad joint.

With any luck it will now be possible to remove the pivot pin; should it be a little tight, hold it in a vice and give the cylinder block a quick twist, and the pivot pin should then unscrew without any further bother, providing the graphite has been applied properly. All that now remains is to continue the hole that acts as a port, from the port face right through into the bore of the cylinder, and to clean the job. Rubbing the port face in a figure-of-eight motion on a sheet of very fine abrasive material until it is thoroughly clean and completely flat should clean it.

THE FRAME

Like the cylinder, the column or frame is a fabrication made from brass; the pivot hole should also be marked with a centre punch, using a line scribed centrally along the frame, and the dividers can again be used to scribe an ark for the ports. This time the arc should be scribed right across the metal, and the tiniest of

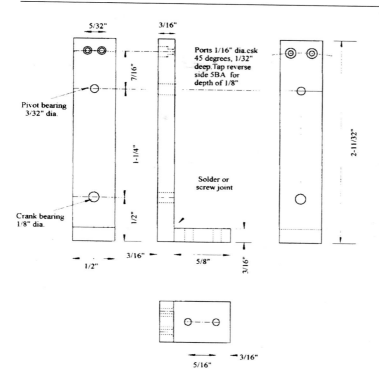

5/32"   3/16"

Ports 1/16" dia.csk
45 degrees, 1/32"
deep.Tap reverse
side 5BA for
depth of 1/8"

7/16"

Pivot bearing
3/32" dia.

2-11/32"

1-1/4"

Solder or
screw joint

Crank bearing
1/8" dia.

1/2"

3/16"

1/2"   5/8"   3/16"

5/16"   3/16"

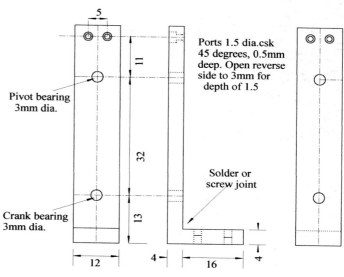

5

Ports 1.5 dia.csk
45 degrees, 0.5mm
deep. Open reverse
side to 3mm for
depth of 1.5

11

Pivot bearing
3mm dia.

32

Solder or
screw joint

13

Crank bearing
3mm dia.

12   4   16   4

Frame: fabricate from brass.
See text for alternative bases.
ABOVE: imperial measurements;
RIGHT: metric measurements.

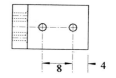

8   4

A centre line is drawn down the port or steam block, which we might also refer to as the frame; in this case a pair of odd leg callipers are being used.

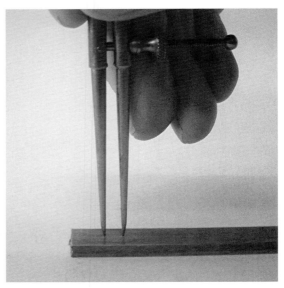

Ideally a jig should be used to drill the steam ports; however, anyone not wishing to go to the trouble of making a jig can make a reasonably accurate job using a pair of dividers. Obviously if working in this way the ports must be marked off from the centre punch mark for the pivot – it is too late to do so once that has been drilled.

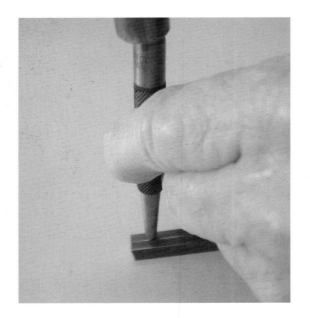

The pivot and bearing holes are lightly centre-punched prior to drilling.

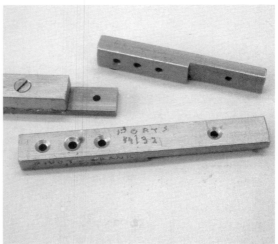

Using a jig such as the one in the photograph for drilling the steam port; it is easy to make, and its use ensures accuracy.

centre-punch marks should be made on the centre line at the intersection. From that mark, it is possible to scribe arcs and locate the position of the ports on the arc, with the dividers. After gently centre-punching them, they can be drilled and lightly countersunk in the same way that was done with the cylinder block.

While the method explained above *will* work, a far more accurate method is to make a jig, which takes just a few minutes to construct. In its most basic form it consists of only a strip of steel with three holes at the spacing of the port and pivot, plus one at the position where the crankshaft will be. The top hole is drilled to the size of the steam port, the next a clearance size for the pivot, and the third a clearance size for the crank pin. The device can only be used after construction of the crank, and then the appropriate hole is put over the crank pin, which is set at 90 degrees. A piece of metal the diameter of the pivot is pushed through the pivot hole in the frame and the jig, and by passing a drill through the hole in the jig representing the steam port, this will transfer a hole to exactly the right position. Set the crank to the other 90-degree position, and there you are, with two perfectly placed steam ports. It is far more accurate than the marking out method.

A slightly more sophisticated version of the jig can also be made very quickly; in most respects it is the same as the one described above, the difference being that another strip of metal is fixed over the end where the pivot and port are. This metal should be the same thickness as the crank; this now gives a much firmer base for drilling the holes as the end is fully supported.

The Base

A base is needed in order that the engine can stand up, and there are a number of methods of making this. The drawing shows a length of the same material as that used to make the frame soldered on one side. A hole must be drilled so that the assembly can be mounted on a block of

A photograph of the column or frame after drilling has been completed.

wood or something similar. This is a quite satisfactory arrangement if the engine is to be screwed down, and it also has the advantage of being the most simple. If the engine is not to be screwed down, or if it is desirable to give extra support, the base should be extended on either side of the frame – in which case it makes sense to solder the column on top of the base, rather than at the side as shown in the original sketch.

To ensure a good mating surface between the cylinder port face and the steam block, both should be worked lengthways along a very fine file. Do not be tempted to put the job in the vice and rub the file on it, as doing so will almost certainly end up with a rocking motion and have the opposite effect of that required.

For a real Rolls-Royce of a job, the more skilled builders may consider removing a piece from the base and fitting the frame into that. It is not all that difficult to do, but it is essential that a good fit be made.

THE CRANK

The crank is used to convert the linear motion of the piston to a rotary one, and the dimensions have a direct relationship with the pivot and ports; it is therefore essential that care is taken to get the measurements as near to correct as possible. The crank in this case only consists of a length of steel bar, with two holes, one for the crank pin and the other for the drive shaft. It is absolutely essential that these two holes are at 90 degrees to the surface of the bar: if they are not, the engine will at the very least have tight spots, and at the worst will not work at all.

Crank Construction

Take a short length of mild steel bar and thoroughly clean it, then spread a marking fluid along one face. Marking fluid can be bought as such, but generally it finds limited use and most gets wasted; a good cheap alternative is to use a paint-marking pen, which can be bought quite cheaply at a whole variety of shops and supermarkets. Old-time engineers did not have the luxury of marking fluids, let alone paint markers, and used a copper sulphate solution. Both paint markers and marking fluid rapidly wear off the work, whereas copper sulphate is

much more permanent and some readers may wish to try using it. To make it, all that is needed are some copper sulphate crystals, obtained from a good chemist or hardware shop: a couple of teaspoons of the crystals are dissolved in water – an instant coffee jar is about the ideal size. The crystals are simply painted on the steel, which promptly takes on a nice copper colour, and any scribing marks show up as silver through it.

A word of warning, though: while marking fluid and paint markers can be applied to a piece of steel that has just been wiped over with a rag soaked in a solvent such as white spirit, to apply copper sulphate the steel needs to be more thoroughly cleaned and should be rubbed with emery paper. Because the solution is water-based, it cannot deal with a surface that has even the slightest trace of grease or oil, and solvents such as white spirit actually leave their own deposit on the metal, which will prevent the copper sulphate from acting.

Having coated the mild steel bar, use a pair of odd leg callipers (Jennies) to scribe a line exactly down the centre. Make a small indentation on the line with a centre punch, and then use a pair of dividers to scribe an ark where the second hole is to be made, and lightly centre punch where the scribed line intersects with the longitudinal one. Drill the necessary holes ensuring the drill is absolutely square, starting with a centre drill and opening out with drills as required. To ensure the drill goes through square, the work should be held in a

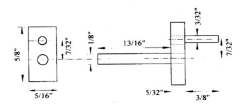 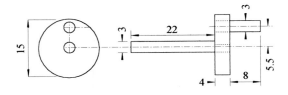

Crank pin and crankshaft: fabricate from mild steel.
LEFT: imperial measurements; RIGHT: metric measurements.

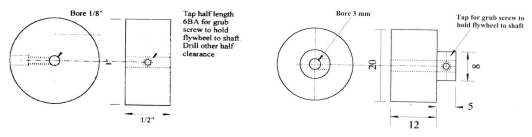

Flywheel: make from mild steel.
LEFT: imperial measurements; RIGHT: metric measurements.

vice that is bolted to the drilling table, and either an engineers' square or a specially made square block used to check the position. The latter is generally preferable, as most engineers' squares will be too large for the purpose.

Note that the end of the crank that is opposite the pin has been left extended as a counterbalance; this ensures a smoother running engine, and also assists the flywheel in maintaining the momentum. There are a number of ways to fit the shaft and pin to the crank, but the following two are recommended: drill and tap and thread the both to fit; or make them a push fit and secure with a retaining compound. When finally assembling the engine, a washer should be placed between the crank and the column in order to reduce friction.

THE FLYWHEEL

Making the flywheel is simply a case of drilling a short length of mild steel bar and parting it to length. The width of the wheel is not important, except that so that the engine can be dismantled if required, it will have to be held in place with a grub screw. This involves cross-drilling the wheel and tapping it. Getting a drill central when cross-drilling is always a little difficult, but by laying a short length of thin mild steel strip (a steel ruler will do) and bringing a centre drill to touch it, the exact position can be seen as the strip or rule will

deflect downwards towards the side unless it is central. If we adjust the position of the work until the piece is at 90 degrees, then we know that the centre of the drill chuck is exactly above the centre of the work. Because of the depth involved there is no point in tapping the whole length of the hole, and so a clearance drill should be run to about half the depth, and the rest can be tapped.

One problem with holding a component in place with a grub screw is the tendency of the screw to throw up burrs and in doing so making it difficult and sometimes even impossible to remove the piece. A small flat filed on the shaft for the screw to locate on will avoid this problem.

PISTON, CONNECTING ROD AND BIG END

In order to make life easy for the less experienced, a simple option has been taken for the above components by combining them all in one. Some people may decide the arrangement looks wrong, and if so, later chapters describe other methods that can be adapted. All that happens for this engine is to cut a suitable length of brass or stainless-steel rod, and cross-drill a hole for the crank pin. The rod acts as piston, piston rod and big end all in one, and saves a great deal of time. It is also an easy means of lining things up with accuracy.

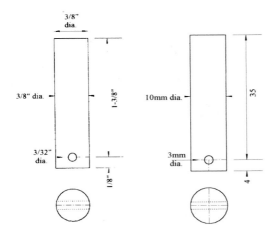

Combined piston and piston rod: one off brass or stainless steel. LEFT: imperial measurements; RIGHT: metric measurements.

STEAM CONNECTIONS

Things are nearing completion, but while we have the necessary steam and exhaust ports, so far there is no way that we can connect a steam pipe to the engine. Strictly speaking it is only necessary to make one fitting, allowing the exhaust to escape through the other port; whether to leave it like that, or to make two connections will be a choice for the individual builder.

The actual connecting piece should not present any difficulty, although care must be taken to ensure that the reduced section that fits the engine frame is reasonably accurate. The fitting is to be soft soldered to the frame or column, and in order to do this it is necessary to open out the port, or ports, to accept it. While theoretically this is just a case of putting a larger drill through the back, less experienced readers should be very wary. When doing this on brass there is always a tendency for the work to snatch, and try and climb up the drill; if it does so the work will be completely ruined. Two things are required to avoid this: firstly the work must be securely held down, either with clamps or in a vice; and secondly a sharp drill

must not be used under any circumstances, and to avoid this, the cutting edges of the drill being used should be rubbed over with an oilstone to ensure it is blunt.

Soldering the fitting or fittings in place follows the usual pattern of ensuring cleanliness, fluxing and heating and is straightforward enough. If the application of solder is overdone, some may run on to the port face, in which case it should be removed with fine emery paper or cloth, laying the latter on a known flat surface and rubbing the engine frame in a figure-of-eight motion on it until all traces of solder have gone. Anyone wishing to connect the engine permanently to a boiler or manifold can solder the pipe straight into the frame.

FINAL ASSEMBLY

Assembling the engine is straightforward, the only additional part required being a small spring, which can be wound using the

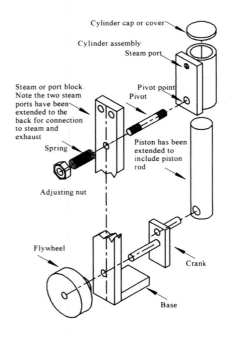

Drawing showing assembly of engine.

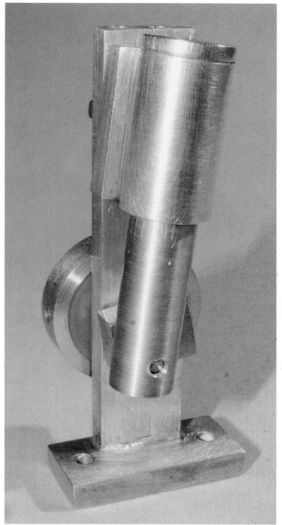

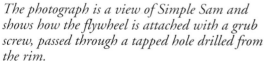

The photograph is a view of Simple Sam and shows how the flywheel is attached with a grub screw, passed through a tapped hole drilled from the rim.

A view of Simple Sam, in which the combined piston and piston rod arrangement can be clearly seen.

lathe, or alternatively a commercial one could be used. The engine should be free to move when the flywheel is tweaked, but at the same time there must not be too much slop. By adjusting the nut that holds the spring in place a suitable tension will soon

be found. In common with engines of any type, the oscillator improves as it is run in; there are almost certain to be high spots in the bore of the cylinder and a slight unevenness between the port faces, which will quickly wear into place.

4 Tin Can Tommy

This is also a simple, basic, single-acting oscillating engine that can be made without the use of a lathe, and – for the benefit of readers not wishing to venture into boiler making until they have gained a little more confidence – an ordinary steel can of the type used for preserving food has been pressed into service. It should be stressed right from the beginning that it is not the best type of boiler and has many limitations, nevertheless the engine will work, although running time will be less than if a proper boiler is used. However, having made the model and thus learned how these things work, it will be a very simple matter at a later stage to transfer the engine to a boiler of a more substantial design.

The first part of the construction is identical to that of Simple Sam, and exactly the same methods are used, but there is no base needed. As the engine is to be fixed directly to the boiler, a short length of the same material that has been used for the port block has to be soldered to the end where the ports are situated. This accepts the length of tube that connects the engine to the boiler and has to be drilled to accommodate it. (A similar arrangement is used for the engine in Chapter 7, and more details are given of the work involved.) The ideal arrangement is to tap the hole and screw the tube in place, but if taps and dies are a problem, the tube can be soldered in place. This will need considerable care to prevent the additional piece falling off the port block. The hole for the tube must just break into the steam port, and it is necessary to ensure that when the transfer tubing is screwed in place, it does not cover the link between the hole and steam port. Likewise if it is soldered it will be essential to ensure the solder does not encroach in the port.

A second, smaller hole is drilled opposite, and this, too, must just break into the ports; it is there to provide an escape for exhaust, and unless one wishes to fit an exhaust pipe, there is no need for that hole to be tapped.

THE BOILER

Well, it's hardly a boiler: it is simply a tin can with two holes drilled in the top, one which accepts a filler cap, the other the connection for the engine. In both instances if nothing else is available, brass nuts can be used. Apart from the obvious fact that the one used for the connection must fit the connecting tube, almost any size of nut will do. The second opening is for filling the boiler, and a screw will be needed to seal it after filling. This must be capable of being screwed down so that it is

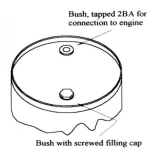

Bush, tapped 2BA for connection to engine

Bush with screwed filling cap

Scrap view showing arrangement on top of can.

A view of the set-up for mounting the engine on the can.

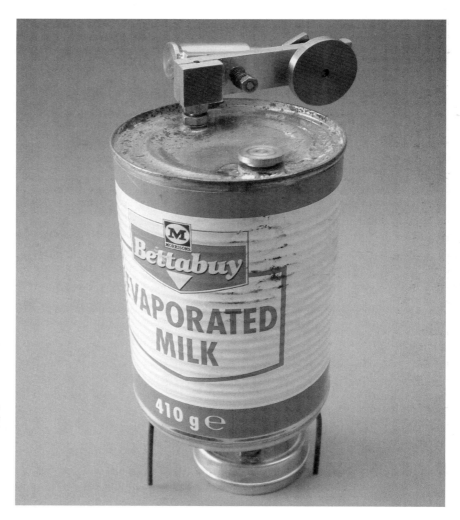

Showing the complete set-up, with the stand that was made from a wire coat-hanger, and the burner made from a lozenge tin underneath the can.

Soft solder brass bush with 1/4" (6mm) dia, bore, to lid as wick holder.

Burner made from lozenge tin with lift-off type lid.

A close view of the lozenge-can burner.

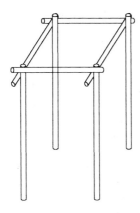

Arrangement for can support. Make from stout wire to fit inside bottom lip of can and solder all joints.

steam tight; an 'O' ring underneath the head will help.

If, because of lack of threading and tapping facilities, it is necessary to solder the engine connection tube in place, it will be best to leave the bush off the can that is used as a boiler, and to solder the tube straight into the can. Trying to soft solder the tube into a bush that has already been soft soldered in place is inevitably going to result in a bad joint on one of them.

THE BURNER

Again, it is a case of simplicity: an ordinary small lozenge tin has a hole drilled in the centre, and a length of copper tube slipped in it and soldered in place: and that is it. In use the lid is removed, and the fuel put in the tin, which is closed and can then be lit for use.

The Boiler Stand

The stand was made from an old wire coat-hanger that was cut up, bent to shape and

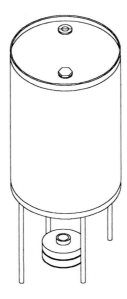

General view of method of arranging engine and food can.

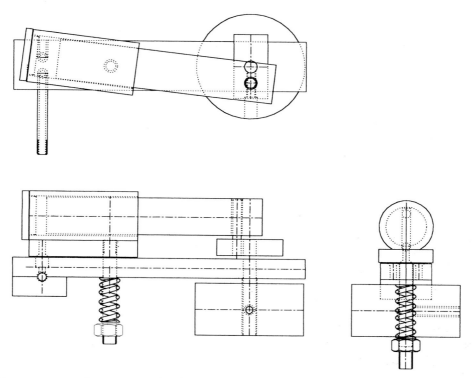

A general arrangement showing completed assembly.

soldered together. It should be made to fit inside the lip of the boiler to prevent the latter falling off. The height of the stand will depend on the height of the flame from the burner, and so no measurement can be given; it is therefore better to make it longer than required, and reduce the height as needs be.

OPERATING

Readers will have noticed that there is no control valve: this is for safety reasons, as it might be possible for such a valve to be closed down to the point where the boiler could explode. It is unlikely, but we must guard against such a possibility. Before running the engine it is essential to ensure that it will run very freely, as the steam supply will be at a low pressure and very limited. Of course it is always necessary to make sure an engine is operating well before it is connected, and in this case it is even more important to do so. If an engine is really well made and designed it should be possible to blow into it with the mouth and make it turn over.

Before running the engine, put a couple of drops of oil on the bearings and in between the port face and port block. To run it, put some water in the boiler: at the very most it can be about a quarter full, but to start with, less is recommended. The engine should be set with the crank at either top or bottom dead centre and the burner lit. When the water is heard to be boiling, give the flywheel a twist and the engine should start to run. And if it does not, put it back to dead centre again and allow more steam to build up. It is a matter of practice making perfect.

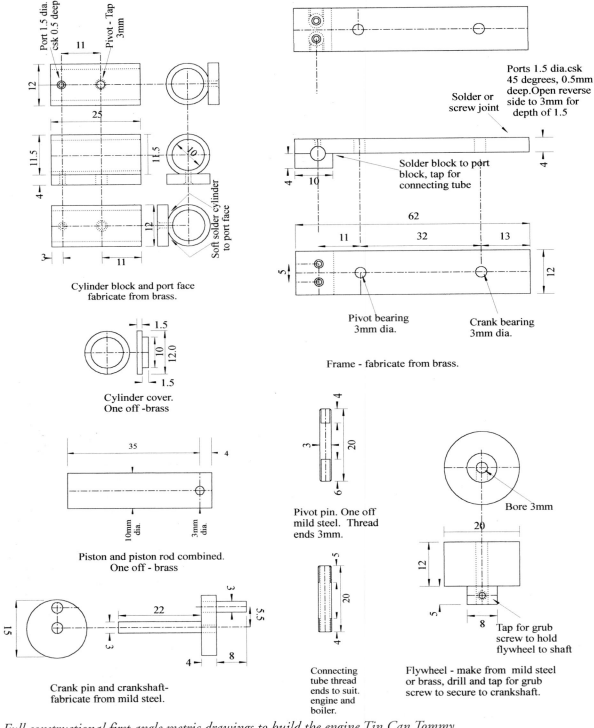

Port 1.5 dia. csk 0.5 deep

Pivot - Tap 3mm

Soft solder cylinder to port face

Cylinder block and port face
fabricate from brass.

Cylinder cover.
One off -brass

Piston and piston rod combined.
One off - brass

10mm dia.

3mm dia.

Crank pin and crankshaft-
fabricate from mild steel.

Ports 1.5 dia.csk
45 degrees, 0.5mm
deep.Open reverse
side to 3mm for
depth of 1.5

Solder or
screw joint

Solder block to port
block, tap for
connecting tube

Pivot bearing
3mm dia.

Crank bearing
3mm dia.

Frame - fabricate from brass.

Pivot pin. One off
mild steel. Thread
ends 3mm.

Connecting
tube thread
ends to suit.
engine and
boiler.

Bore 3mm

Tap for grub
screw to hold
flywheel to shaft

Flywheel - make from mild steel
or brass, drill and tap for grub
screw to secure to crankshaft.

Full constructional first angle metric drawings to build the engine Tin Can Tommy.

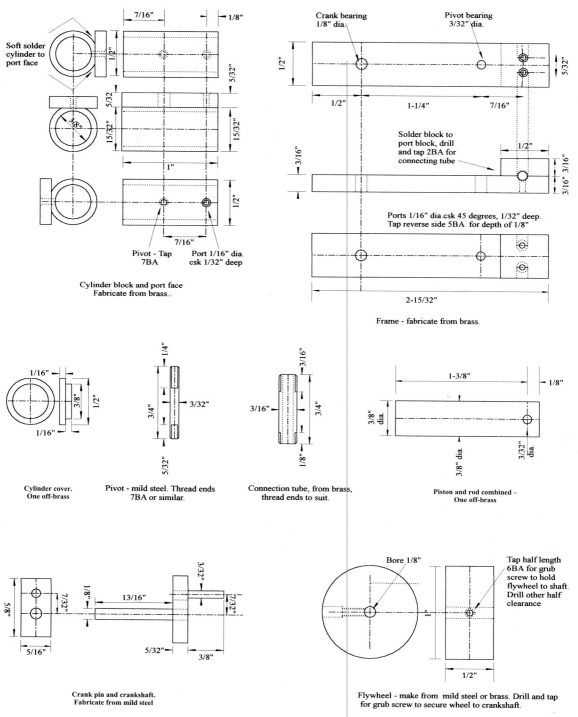

Soft solder
cylinder to
port face

7/16"
1/8"
1/2"
5/32

5/32
15/32"
15/32
1"

1/2"
7/16"

Pivot - Tap
7BA

Port 1/16" dia.
csk 1/32" deep

Cylinder block and port face
Fabricate from brass..

Crank bearing
1/8" dia.

Pivot bearing
3/32" dia.

1/2"
5/32"
1/2"
1-1/4"
7/16"

3/16"

Solder block to
port block, drill
and tap 2BA for
connecting tube

1/2"
3/16"
3/16" 3/16"

Ports 1/16" dia.csk 45 degrees, 1/32" deep.
Tap reverse side 5BA for depth of 1/8"

2-15/32"

Frame - fabricate from brass.

1/16"
3/8"
1/2"
1/16"

Cylinder cover.
One off-brass

1/4"
3/4"
3/32"
5/32"

Pivot - mild steel. Thread ends
7BA or similar.

3/16"
3/16"
3/4"
1/8"

Connection tube, from brass,
thread ends to suit.

1-3/8"
1/8"
3/8" dia.
3/8" dia.
3/32" dia.

Piston and rod combined –
One off-brass

5/8"
7/32"
1/8"
13/16"
3/32"
3/32"
5/16"
5/32"
3/8"

Crank pin and crankshaft.
Fabricate from mild steel

Bore 1/8"

Tap half length
6BA for grub
screw to hold
flywheel to shaft.
Drill other half
clearance

1"
1/2"

Flywheel - make from mild steel or brass. Drill and tap
for grub screw to secure wheel to crankshaft.

Full set of imperial constructional drawings for Tin Can Tommy.

5 Slim Sam

This engine is an upgrade on Simple Sam; there are some different measurements, but the basic difference is in the fact that the solid piston/big end configuration is replaced with a much neater stainless-steel piston rod and a big end bearing, and while the latter is not quite in keeping with full-size appearance, it does have a very similar look to it, enough anyway to fool the uninitiated into thinking it is a scaled-down version of the real thing. It is still a single-acting single-cylinder engine, but the new configuration should give slightly better running qualities, as there is less heat loss than was the case with the solid combined piston. The alterations give it a slimmer, less chunky appearance than Simple Sam.

A slightly wider distance between the ports gives more swing to the cylinder, and this works

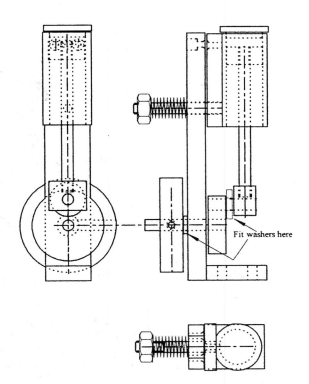

General view of 'Slim Sam', showing layout and proportions.

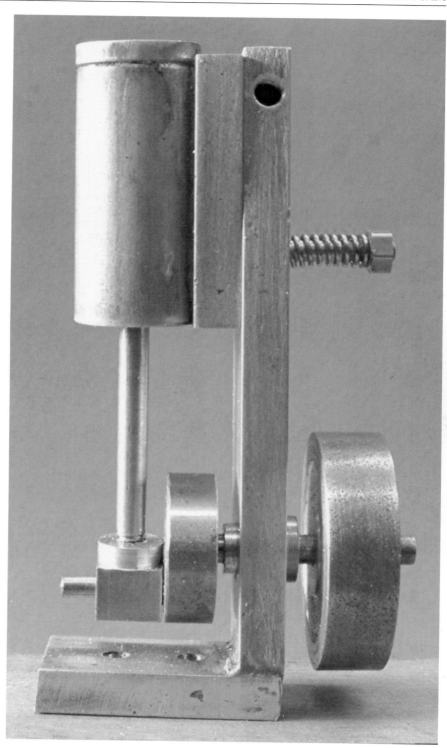

A general side view of the engine, showing the slimmer and possibly more business-like look than that of Simple Sam.

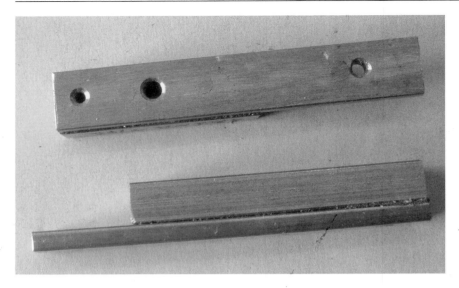

Two examples at different angles of the jigs used for marking out port positions, and as used on this engine.

better with the piston rod/big end arrangement; the extra distance between the ports normally means an engine that is more difficult to start and that operates at a faster speed, but this is compensated for by countersinking the cylinder port slightly as well as those on the steam block, allowing an earlier entry and exit of steam than would otherwise be the case. As a result of these differences the engine is a little larger than its smaller brother Sam; it also has a more elegant look, but still in keeping with the general theme of the book. Also in keeping with the principles of the book is the fact that the engine is just slightly more difficult to build, with a small amount of extra work.

CYLINDER AND FRAME

Manufacture of these all follow the pattern that has already been explained in depth, and basically the same materials have been used so further explanation should be unnecessary. If a spacing jig is being used, a different one will be required in order to make the engine as per the drawings; but should readers wish to, they can use the same measurements as those of Simple Sam, and adapt the piston rod and bearing to suit.

THE PISTON

The piston can be made of brass, bronze or stainless steel. In theory the latter is the best, as running two identical or near-identical metals against each other leads to rapid wear. Having said that, it is most unlikely that any of these engines will ever get so much usage that wear will be a problem – not, that is, unless they are handed down from father to son for several generations, and then it could happen. So while stainless steel is the correct material to use, it is very hard and is most difficult to work with, so readers must make their own choice.

It is always best to machine an oversized piece of metal in order that the diameter is a good fit in the cylinder; machining the outside, and drilling and tapping at one setting of the machine also ensures absolute concentricity. Should stock material be used instead, then in all probability it will require lapping to get a smooth movement in the bore. In this case it can be done by screwing in the piston rod, spreading a layer of metal polish over the piston, and working it in and out of the bore until a smooth operation is felt. It might be necessary to recharge the metal polish a few times when doing so, and both the bore and piston surface must be thoroughly cleaned afterwards, using a solvent such as white spirit.

Even when using stock material, the drilling and tapping of the piston is best done in a lathe, in order to achieve the maximum accuracy. This does not mean that the engine cannot be made by someone who does not have access to a machine, as with care, the piston could be made by hand.

The Piston Rod

The piston rod should definitely be stainless steel. Brass is not really suitable, and if mild steel is used it is likely to become rusty in a short period of time, because inevitably there will be some steam leaking from between the cylinder walls and the piston. All that is needed for the piston rod is to thread both ends of a short length of material of the correct diameter; again, it is a job best done with the use of a lathe, but can be managed without if necessary. Getting small threads square when using a die-holder by hand is far from easy and likely to lead to the piston not being perfectly square on the rod; so if it is to be done that way, take extra care to ensure the die does go on the material correctly.

BIG END BEARING

A short length of square brass or bronze bar is required for the big end bearing; it has to be cross-drilled for the crank pin, and then drilled and tapped lengthways in order that the piston

rod can be screwed in. This, too, is really a job for the lathe, in which case it has to be done using a four-jaw independent chuck to hold the material. A rather nice finish to the bearing can be obtained if at the same time a short circular section is machined on the end around the thread, but that is only a matter of choice, and doing so will in no way improve the running of the engine.

If the bearing is made by hand, take care to ensure the thread is running centrally into the bar stock, and that the hole for the crank is at 90 degrees to it. Should the two not be at 90 degrees it is very doubtful whether the engine will work.

THE CRANK SHAFT

Although to some extent a cosmetic change, the shape of the crank has also been altered, with a disk being used in place of the more usual rectangular-section material. This is not uncommon on full-size engines, and the idea is to give a better balance to the engine. It is doubtful if there is any such advantage with an engine of this size, but this type of crank looks attractive when the engine is running, particularly if it is painted in a bright colour. The fact that a round crank is used does not mean the crank has to be made in a lathe, although obviously doing so is best. Provided care is taken to keep both the crank and the shaft holes square, it can be drilled in a drilling

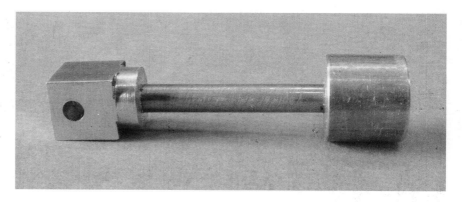

The piston, piston rod and bearing set-up used on the engine; while it involves more work than the arrangement of Simple Sam, the appearance is much improved.

machine. The fact that the crankshaft hole might be a fraction off centre will not make the slightest bit of difference.

Builders using a lathe will find it best if a four-jaw chuck is used to get the offset hole for the crank. Failing that, the central one can be made in the lathe, and the crank then taken to the drilling machine to make the second one.

FLYWHEEL

The flywheel follows the usual pattern, and can be made of brass or steel; final details are left for the builder to decide, whether to use the one shown on the drawings or to select one from another engine. As usual, a grub screw is the best way to secure the wheel to the shaft, although there are some people who like to thread the shaft end and make a mating thread in the wheel so that it can be screwed in position.

It will be noticed that the steam connections have been brought out of the side of the steam block on the model in the photographs, as well as shown in that fashion in the drawings. In common with all the oscillating engines, this is arbitrary, and the builder can reposition them according to his or her wishes depending on the boiler to be used and the position the engine will occupy in the final set-up.

This photograph clearly shows the set-up of the piston rod and big end.

A view from the other side, showing the arrangement for the grub screw to secure the flywheel.

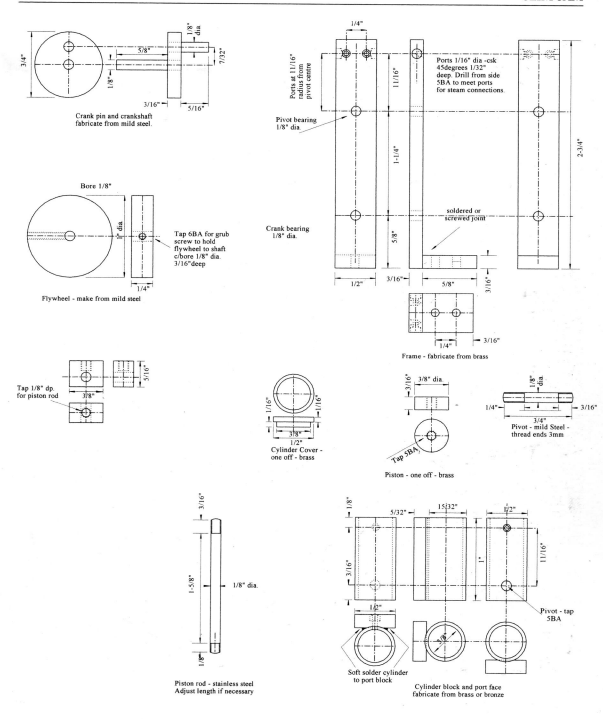

Crank pin and crankshaft
fabricate from mild steel.

Bore 1/8"

Tap 6BA for grub
screw to hold
flywheel to shaft
c/bore 1/8" dia.
3/16"deep

Flywheel - make from mild steel

Ports at 11/16"
radius from
pivot centre

Ports 1/16" dia -csk
45degrees 1/32"
deep. Drill from side
5BA to meet ports
for steam connections.

Pivot bearing
1/8" dia.

Crank bearing
1/8" dia.

soldered or
screwed joint

Frame - fabricate from brass

Tap 1/8" dp.
for piston rod

Cylinder Cover -
one off - brass

Tap 5BA

Piston - one off - brass

Pivot - mild Steel -
thread ends 3mm

Piston rod - stainless steel
Adjust length if necessary

Pivot - tap
5BA

Soft solder cylinder
to port block

Cylinder block and port face
fabricate from brass or bronze

Slim Sam. Imperial version.

53

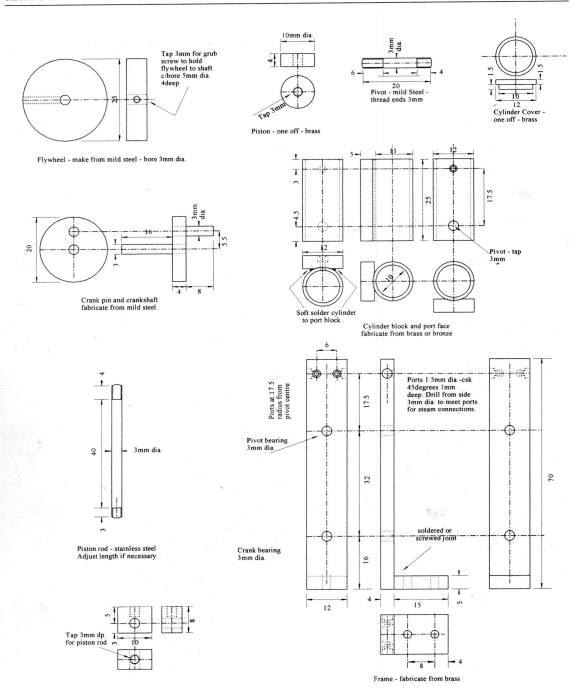

Tap 3mm for grub screw to hold flywheel to shaft c/bore 5mm dia. 4deep

Flywheel - make from mild steel - bore 3mm dia.

10mm dia.

Tap 3mm

Piston - one off - brass

Pivot - mild Steel - thread ends 3mm

Cylinder Cover - one off - brass

Crank pin and crankshaft fabricate from mild steel.

Soft solder cylinder to port block

Pivot - tap 3mm

Cylinder block and port face fabricate from brass or bronze

3mm dia.

Piston rod - stainless steel
Adjust length if necessary

Crank bearing 3mm dia.

Ports at 17.5 radius from pivot centre

Pivot bearing 3mm dia.

Ports 1.5mm dia -csk 45degrees 1mm deep. Drill from side 3mm dia. to meet ports for steam connections.

soldered or screwed joint

Tap 3mm dp. for piston rod

Frame - fabricate from brass

Slim Sam. Metric version.

54

6 Millie, a Mill-Type Engine

The horizontal steam engine was used extensively in industry throughout the world, and it was to be found powering machinery of all types, with the result that such engines were manufactured in an extensive range of sizes; those used to power the woollen and cotton mills of Yorkshire and Lancashire were absolutely massive. It is not proposed to build a model of one of those massive brutes, but the horizontal engine can be nicely simulated as an oscillating engine, and it makes a most attractive model. The major difference in this engine is the fact that rather than being built around the port block, the engine is built on a base.

THE CYLINDER

The cylinder and port face assembly of the engine follows a similar pattern to the other engines described in this book, but in keeping with the full-size prototype, the cylinder is

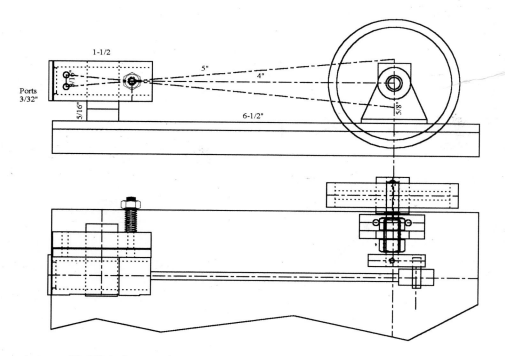

General view of 'Millie', showing layout and proportions.

View of the engine showing in particular the method of supporting the crank and the fish-bellied connecting rod.

The other side of the engine; the cylinder support and steam connections can clearly be seen.

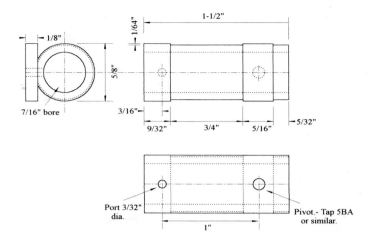

Cylinder arrangement for Millie.

larger. In the interests of simplicity, the engine has been made single-acting. Manufacture can initially therefore follow the same method as with previous oscillating engines.

PORT BLOCK

The first major difference, as already mentioned, is in the port block, which in this model does not have a facility for supporting the crankshaft; that is placed on a separate bearing bolted to the base. We only need therefore to make the steam ports, and provision for the pivot pin. These are slightly larger than those on the previous engines, but otherwise there is no difference. The port block is fixed to the base via a stand which is fabricated from the same brass strip as the block itself; the example shown in the photographs is screwed together, and this is reflected on the drawings. In many ways it would be easier to soft solder them and there is no objection to that idea, other than the fact that it can be more difficult to get everything square when soft soldering. The choice of how to make it is therefore left to the reader.

The stand is also shown as screwed to the port block, the screw holes passing right

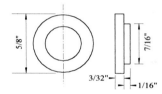

Cylinder cover: one off – brass. Soft solder to cylinder.

through the block. This will in no way affect the running of the engine, as long as they are thoroughly de-burred, but it does make manufacture much easier as there are no blind holes to tap. Again, if readers would prefer to soft solder the bracket to the port block there is no reason why not, as long as it can be assured that the edge of the block is parallel to the base.

PISTON AND PISTON ROD

The pivot pin is the normal type, and the cylinder is secured with a spring and nut. The piston and piston rod plus the big end bearing follow the pattern of manufacture used in the last chapter, with the rod threaded into both piston and bearing. The drawings show the rod

57

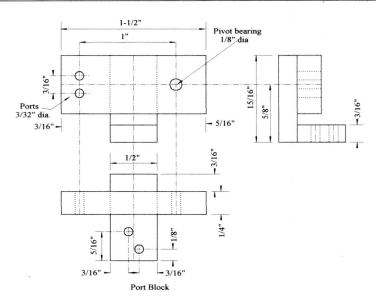

Port block and support: one off – fabricate from brass.

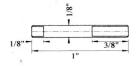

Pivot pin: one off – stainless steel. Thread ends 5BA or similar.

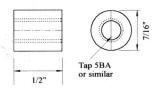

Piston: one off – brass.

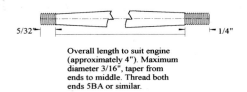

Overall length to suit engine (approximately 4"). Maximum diameter 3/16", taper from ends to middle. Thread both ends 5BA or similar.

Connecting rod: one off – mild or silver steel.

as parallel throughout its length, but it was quite common for them to be fish-bellied – that is, of a larger diameter in the centre than at the ends; for appearance sake this would be preferable, but it makes no difference whatever to the operation of the engine.

For the benefit of anyone proposing to make a fish-bellied rod, here are a few ideas. Start by machining the ends and threading them, then machine a taper on one end about three-eighths of the length of the rod, discounting the thread length. Turn the work around and repeat the operation at the other end of the rod. Put a nut that will accept the thread on the end of the rod in the three-jaw chuck, or better still a collet if you have one, and screw the rod into it. Put a fixed steady at the other end just short of the thread, and adjust it to ensure the rod is running true. Even if the original machining is not completely accurate, it should be possible to make suitable adjustments by using the arms of the steady. With the lathe running at a very high speed, blend the curve to the parallel section, using a very fine file.

Take great care during this operation: there should be a guard on the chuck, and sleeves must be secured with plastic bands to prevent them getting caught in any part of the lathe. It is not an operation recommended in textbooks, but for most home workshops there is no alternative.

Do not worry if there is a short length of flat left, as it will not notice at all when the model is finished.

Finally remove the rod and work on it with varying grades of emery cloth or paper to remove any marks and perhaps enhance the blending a little more. The movement of the abrasive should be along the length of the rod, and not around the circumference; it is a common mistake to do it that way because it gives a shinier finish, but it will prove near impossible to get the transition from straight to taper right. Finally polish the rod using a metal polish and cloth; how much of a polish is imparted is a matter of personal choice, but in full size, although rods were kept bright and clean, it was very rare to see them with a high polish.

CRANK

The crank is made as before, using either steel strip or round section, according to preference. Most cranks were built up with a weight on the side opposite the crank pin, and a number of companies preferred to use a circular crank. The crank pin can either be screwed in place or secured with a retaining compound; some may even wish to silver solder it. In theory there should always be a nut of some sort outside the bearing to ensure there were no accidents with it moving off the pin; on a model such as this it is not a problem, and therefore a matter of personal choice whether or not one wishes to put a nut on.

CRANK SHAFT

The crank shaft runs in a bearing, which in turn is in a housing mounted on the base; this can be made from a short length of angle iron

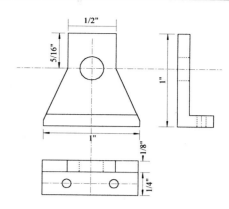

Main bearing housing: one off – make from angle iron.

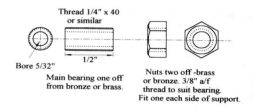

Make up of main bearing for Millie.

with a separate bronze or brass bearing; it is shown on the drawings as made from hexagon material, but some readers may prefer to use round instead. Hexagon was not all that commonly used in full-size practice, although strangely it was not at all unusual to find the outside of them from octagonal material.

The big end bearing is a plain brass block. A small circular section machined where the connecting rod screws in would no doubt improve its appearance.

FLYWHEELS

A feature of any horizontal engine was the large flywheel, and the drawings show a flywheel 3" (7.6cm) in diameter, which is about right for an

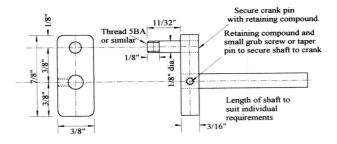

Crankshaft assembly.

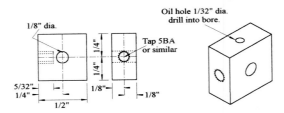

Big end bearing: one off – bronze or brass.

engine of these proportions. However, in the case of an oscillating engine, if it was cast iron or bronze or a similar material it would be far too heavy, and although giving momentum, would also absorb power. The wheel seen in the photographs has been cut from a piece of wood and has a card rim; the central boss is made from steel and does the actual work. Cheating perhaps, but at least it looks the part, it is cheap and easy to make, and in addition is just about the weight we need. Also shown are drawings of other flywheel arrangements that, although less common than the spoked variety, were nevertheless to be seen on engines.

The steam and exhaust connections are taken through the port block; when doing this, the drill should always be set to the side of the ports so that it is just the edge of the hole that breaks in. In this way the connections are wide enough apart for the necessary pipes to fit without fouling each other.

The model should be screwed to a steel or alloy plate, and that in turn mounted on a piece of wood; ideally hard wood should be used if it is available. In the case of the model that is shown, the screws that hold it to the base have passed right through the metal, and recesses made in the wood to accept them. This was done purely to avoid shortening some over-length screws, and it is entirely an individual choice.

The model is ideal for including further embellishments. For example, one might wish to build simulated brick walls round three sides, to represent an engine house. This in turn can be further enhanced with the sort of bits and pieces one might expect to find on the premises, such as shovels, oilcans, buckets, and so on. Keep in mind that all engine houses were treated with great pride by the staff, and maintained in a clean and tidy condition.

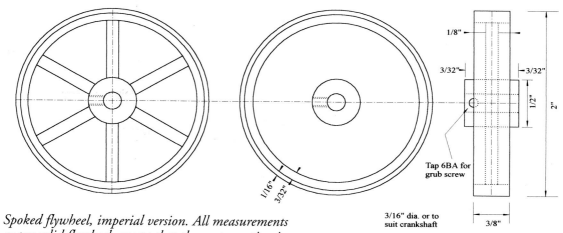

1/8"
3/32"
3/32"
1/2"
2"

1/16"
3/32"

Tap 6BA for
grub screw

3/16" dia. or to
suit crankshaft

3/8"

*Spoked flywheel, imperial version. All measurements
as per solid flywheel, except that the centre section is
removed and replaced with six ⅛" dia. spokes.*

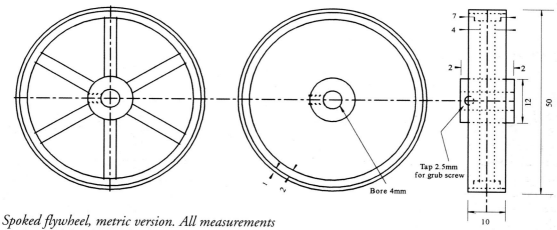

7
4
2
2
12
50

10

Tap 2.5mm
for grub screw

Bore 4mm

*Spoked flywheel, metric version. All measurements
as per solid flywheel, except that the centre section is
removed and replaced with six 3mm dia. spokes.*

*RIGHT: This flywheel has the
advantage of not being too
heavy and it is easier to
make than the spoked type.*

*FAR RIGHT: This flywheel
shape was used from time to
time on pumping engines.*

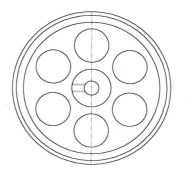
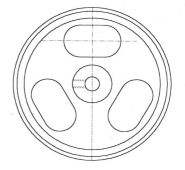

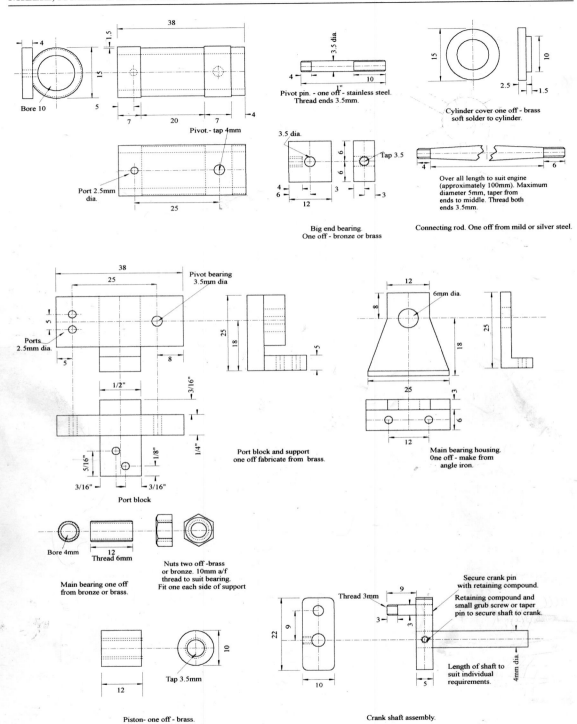

Full set of metric drawings for 'Millie'.

7 Invar, An Inverted Vertical Engine

Full-size engines were built in all sorts of configuration, and generally the way the engine was laid out was because that was the most suitable type for a particular purpose. Very commonly used was a vertical engine that had the cylinder placed at the bottom and the crankshaft at the top: this was known as the inverted vertical type. The layout was very suitable where overhead line shafts were in operation, as it meant that a shorter drive belt could be used than was the case when the shaft was at the bottom or the engine was horizontal. Belting was frequently a problem in all types of establishment, because over a period it would stretch and become slack, and in addition when belts were worn they were prone to snap. All large factories would employ someone whose job it was to go round and ensure the belts on the shafts and machines were in order, and to make any repairs that might be required. In most cases the belt would be joined with brass clips, through which a pin was passed to hold

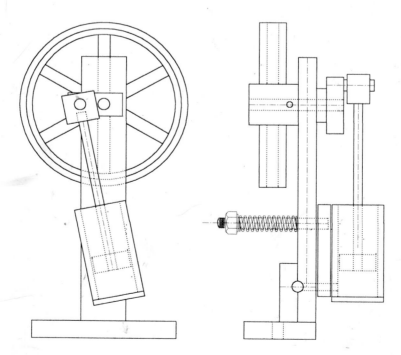

General view of 'Invar'.

the two sections together. It was not unusual for a belt to be made up of several pieces, where in the past it had worn away and a section been fitted in.

This type of engine was generally quite small when compared to the massive great mill engines and therefore would be used in smaller establishments, such as small engineering factories and printing works. Surprisingly they were commonly found driving the limited amount of machinery to be found in breweries as well as in the smaller railway workshops. A feature of the engines was the use of a larger flywheel than those seen on engines with the cylinders at the top, and as a rule the flywheel would be spoked in order to save weight. The

The completed engine; readers will no doubt agree that the spoked flywheel gives it an elegant look, and is well worth the effort of making it. This view shows the cylinder and crank arrangement.

reason for this is almost certainly the fact that the larger diameter meant a smoother action and therefore less vibration at the top of the engine, where an excessive amount could create problems. At the same time, while the momentum on the diameter was needed, the extra weight was not, hence the spokes.

Fabrication of the cylinder and port face assembly follows the usual pattern and by now will present no problems to even the newest recruit to model engineering; but from there on it is all change. Obviously the base is not soldered to the crankshaft end; instead it is fitted just below the cylinder. Because of the position of the cylinder it is not practical to bring the steam and exhaust pipes out of what is normally the top of the body, and to assist with bringing it from the sides and to add stability to the engine, small section has been soldered on and the pipes taken from that. This is similar to the construction of Tin Can Tommy, the difference being that if the extra block on that was a little out of line, it mattered not, but in this case if the block is not soldered squarely in place, it will be impossible to assemble the engine so it is completely vertical.

It is essential that the block is soldered firmly to the body, while at the same time avoiding an excessive quantity of solder, and the best way of doing this is to 'tin' both pieces first and then join them. Tinning involves putting a layer of solder on the parts, then a layer of flux, and finally heating the two so that both lots of solder fuse together. To do this, put plenty of flux on both parts, heat them, and allow the solder to run on them. Before the solder can cool and solidify, wipe it over with a cloth; the result will be a smooth surface with solder, and hopefully not spaces that are bare. The cloth used should be a cotton material: wool will stick to the solder, and man-made fibres will simply melt; thus both will leave a mess that cannot be joined. But even when using cotton material to carry out the operation, it is essential that the solder is wiped away quickly in order to get the right result.

It should go without saying that this is an

To solder the extra section to the body, start by fluxing both well.

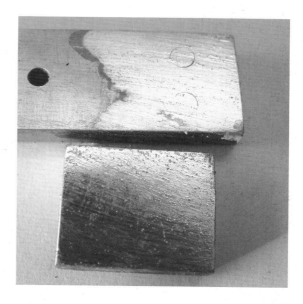

The solder is applied to each, and while still hot, wiped off with either a cotton cloth or a bristle brush. Do not be tempted to use man-made fibres as these will melt and stop the pieces adhering together.

operation that requires a great deal of care, and not one to be carried out over the best carpet. The hot solder will splash off and stick to the floor or any furniture that happens to be around, therefore if possible the operation should be carried out in a garage with a concrete floor or out of doors in a suitable place. Personal care is also necessary to prevent burns, the work should be held in a pair of pliers at the other end from the tinning operation. Gloves must be worn, and again, they should not be made from man-made fibre or wool; good quality cotton garden gloves will do, or it is possible to buy special heat-proof gloves meant for use in the kitchen.

The base also has to be soldered in position, and another advantage of the tinning process shows up here, in that if the area that will mate with the steam block and the end of the steam block itself are also tinned, the parts will join together without any problem. Simply put some more flux on the tinned section, and heat them gently until the solder already in place melts and joins all three together. As a precaution, bind the small piece of brass to the steam block with iron wire just to make certain

that they will not separate during the operation. Anyone with silver soldering facilities can join the steam block to the base with silver solder, and then soft solder the small section in place afterwards.

MAKING THE FLYWHEEL

The rest of the construction other than the flywheel will be found to be easy enough, and if a solid flywheel is used there are no problems. To make the engine look authentic the spoked flywheel is essential; to make it, a piece of tubing will be needed — in the case of this model, the tubing was a slice off a scaffold pole. It happened that the tubing was made of steel, but as a rule it will be found to be of aluminium, which is not ideal for this sort of work. There are all sorts of other possible

To make the flywheel, start by machining a short length of hardwood to be a tight push-fit inside the tubing, and at the same setting drill for the bolt.

It is preferable to tap the wood to accept the bolt, but if a suitable tap is not available the bolt can be passed right through and a nut put on the inside. Make sure the clearance hole is a close fit and the nut is tightened hard to prevent any rotation of the tubing during subsequent operations.

sources of supply for a suitable piece; an obvious one would be a car propshaft cover.

Six evenly spaced holes are needed round the periphery of the tube, and the easiest way to achieve this is to use a large hexagon-headed bolt, something that once again is likely to be available at a scrapyard. Machine a piece of hardwood so that it is a force-fit in the inside of the tube, and, without taking the wood from the lathe, drill a hole for the thread of the bolt to pass through. If it is possible to drill and tap it so the bolt will screw in, so much the better; if not, make sure it is a good tight fit on the bolt. It does not matter whether or not the wood is round, as long as it will fit in the bore of the tube; any surplus can be trimmed off with a chisel. We are not looking for neatness, it is just something to do a job.

With the tube firmly in place and square to the face of the bolt head, it is possible to use each of the flat surfaces of the bolt to drill a hole and so give the required position for six spokes. It is essential to ensure the drill is exactly centrally placed of the circumference of the tube, so start by laying a flat of the bolt head along the edge of the machine vice and tighten up. The poor man's way of ensuring the drill is placed accurately over the tube is to rest a rule on top of the tube and bring a pointed instrument to bear of a rule place on the tube. The pointed instrument is nothing more than a short length of round steel, the point having been machined in the lathe. If the rule is level, then the drill is exactly at the top of the tube, and that is the position at which the drill is passed through. If it is not level, make any necessary adjustments until it is.

This method is accurate enough for our purposes, but for anyone wishing for a more scientific approach a little tool like the one shown in the photograph can be easily made, or can be bought at model engineering suppliers.

All six holes can be drilled by rotating the nut, and then a piece of hexagon bar can be used to obtain the hub of the wheel. The method used is very similar: a short length of bar is machined to the diameter of the hub of

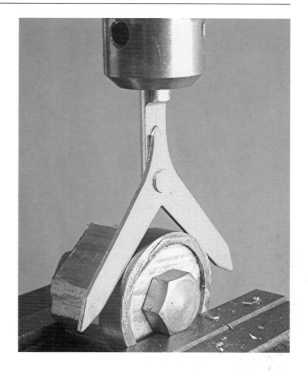

The flats of the bolt are used for dividing, and it is essential that at each drilling operation they are flush along the top of the vice. It is also absolutely essential that the drill is exactly placed at the top of the tube. The device shown here for checking that accuracy can be bought, or is simple to make. It is adjusted until the mark of the vee, and a similar one on the spigot, line up, at which point the drill will be accurately lined up.

the wheel and drilled for the crankshaft of the engine. The holes can be drilled by simply rotating the hexagon bar and using it as a dividing device; when they have been completed, part off the hub.

A simple jig is used for assembling the wheel: it is a piece of suitable hardwood or similar material with recesses bored in it to accept the rim and the hub. Note that these must be accurate enough to prevent movement of either, otherwise the wheel will not run true.

Pass six lengths of mild steel of the appropriate diameter through the rim, into the

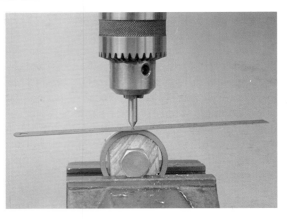

If a special tool is not available, reasonable accuracy for lining up can be obtained with a short rule, or strip of flat metal. A short length of steel that has been machined to a point is placed in the chuck and brought to bear lightly on the metal strip or rule, which will tip to an angle indicating the direction of movement needed.

Here we see the drill has been re-aligned and the rule is now nearer to level, indicating that the drill is very near to the correct alignment.

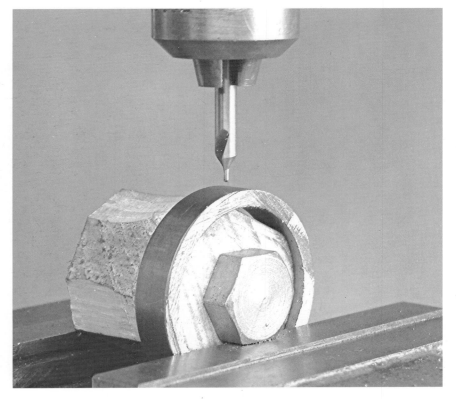

Having found the correct position, start drilling operations with a centre drill; trying to use a drill on the circumference of the tube, without first making a start with the centre drill, will almost inevitably result in the drill moving out of line and making the hole at an angle.

Having successfully drilled six holes accurately in the circumference of the tubing, a jig will be needed to line up the spokes. This need be no more than a piece of chipboard, MDF or hardwood with a recess to accept the tubing. The recess must be accurate so that there is no movement available when the tube is positioned. Provision must also be made for the hub.

The hub is made from a short length of hexagon material; a round section is machined for the actual hub, and then six evenly spaced holes are drilled in it, using once again the six sides of the hexagon material as a dividing device.

hub, checking to make sure the hub is exactly central with the rim and that the spokes are at 90 degrees. They can either be kept in place with a retaining compound, or soft soldered: either will do. Finally when any adhesive has thoroughly set, take a skim of the circumference jut to ensure there are no odd bits of spoke sticking out. Drill the hub for a grub screw, and hey presto! there is the flywheel.

The rim, hub and spokes are all assembled, using either a retaining compound or an epoxy adhesive, and left while the adhesive completely sets.

LEFT: The finished flywheel, which has a rather elegant appearance.

BELOW LEFT: A view of the engine, showing the additional block and the flywheel.

BELOW RIGHT: The opposite side of the vertical engine, showing the position of the cylinder and crank.

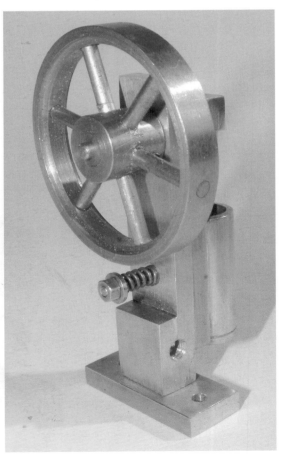

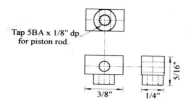

Tap 5BA x 1/8" dp. for piston rod.

5/16"

3/8"   1/4"

Big end bearing - brass or bronze

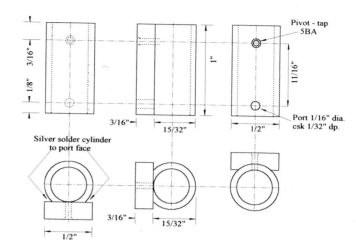

Pivot - tap 5BA

1/8" 3/16"

1"

11/16"

3/16"   15/32"   1/2"

Port 1/16" dia. csk 1/32" dp.

Silver solder cylinder to port face

3/16"   15/32"

1/2"

Cylinder block and port face - fabricate from brass

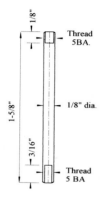

1/8"

Thread 5BA.

1-5/8"

1/8" dia.

3/16"

Thread 5 BA

Piston rod. Adjust length if necessary.

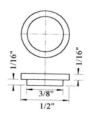

1/16"   1/16"

3/8"

1/2"

Cylinder cover. One off - brass

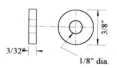

3/32"

3/8"

1/8" dia.

Thrust washer - brass - bore to fit crankshaft. Place washer on shaft between steam block and crank.

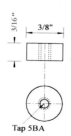

3/16"

3/8"

Tap 5BA

Piston - one off brass or stainless steel

1/2"

Spring - make from bronze or stainless steel wire, 26 gauge.

5BA   1/8" dia.   5BA

5/16"   3/16"

1"

Pivot pin - one off stainless or mild steel.

Drawing for imperial version of inverted vertical engine (1).

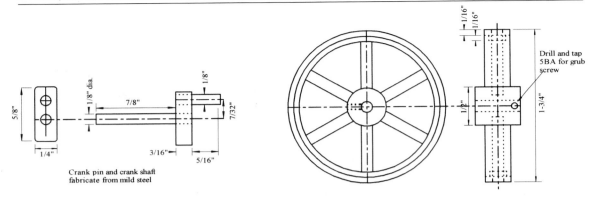

Crank pin and crank shaft
fabricate from mild steel

Flywheel - mild steel. Bore 1/8" dia for crank shaft.
Six spokes 1/8" dia. Fabricate by placing spokes through rim
and into hub. Secure with retaining compound or solder

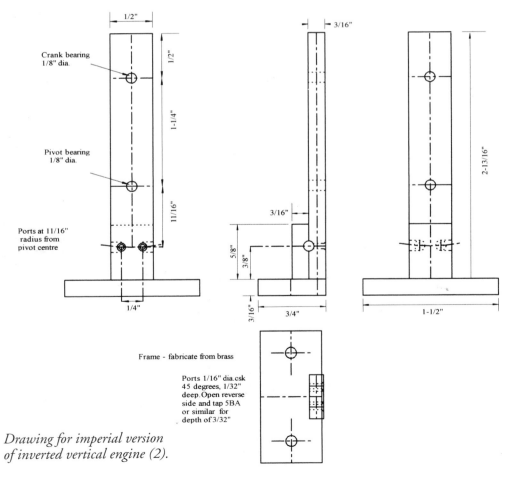

Crank bearing
1/8" dia.

Pivot bearing
1/8" dia.

Ports at 11/16"
radius from
pivot centre

Frame - fabricate from brass

Ports 1/16" dia.csk
45 degrees, 1/32"
deep. Open reverse
side and tap 5BA
or similar for
depth of 3/32"

Drill and tap
5BA for grub
screw

*Drawing for imperial version
of inverted vertical engine (2).*

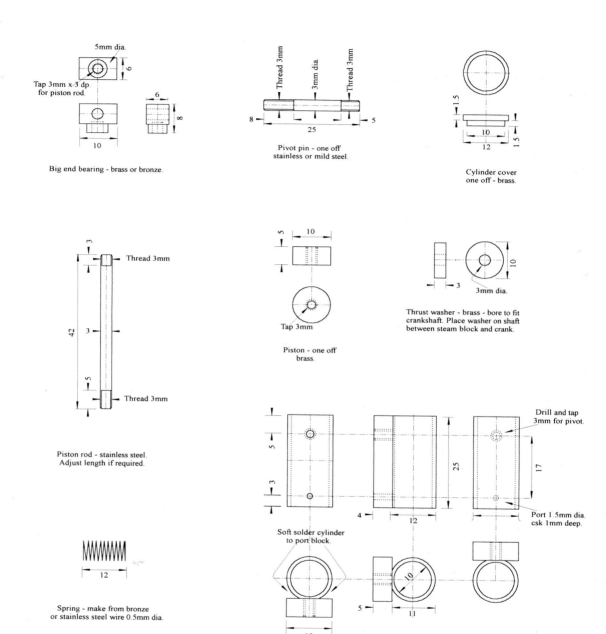

5mm dia.

Tap 3mm x 3 dp.
for piston rod.

6

6

8

10

Big end bearing - brass or bronze.

Thread 3mm

3mm dia.

Thread 3mm

8

5

25

Pivot pin - one off
stainless or mild steel.

1.5

10

12

1.5

Cylinder cover
one off - brass.

3

Thread 3mm

42

3

5

Thread 3mm

Piston rod - stainless steel.
Adjust length if required.

5

10

Tap 3mm

Piston - one off
brass.

3

3mm dia.

10

Thrust washer - brass - bore to fit
crankshaft. Place washer on shaft
between steam block and crank.

12

Spring - make from bronze
or stainless steel wire 0.5mm dia.

Drill and tap
3mm for pivot.

5

3

25

17

4

12

Port 1.5mm dia.
csk 1mm deep.

Soft solder cylinder
to port block.

10

5

11

12

Cylinder block and port face.
Fabricate from brass or bronze.

Metric measurement drawings for inverted vertical engine (1).

73

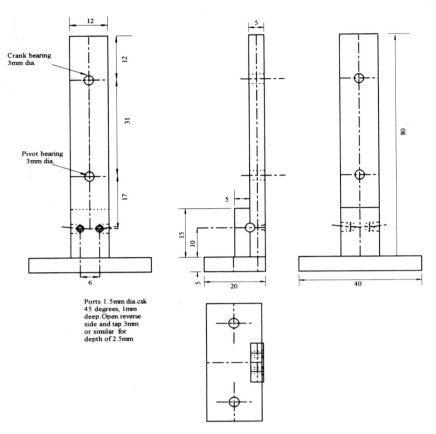

Crank bearing
3mm dia.

Pivot bearing
3mm dia.

Ports 1.5mm dia.csk
45 degrees, 1mm
deep. Open reverse
side and tap 3mm
or similar for
depth of 2.5mm

Frame - metric version, fabricate from brass

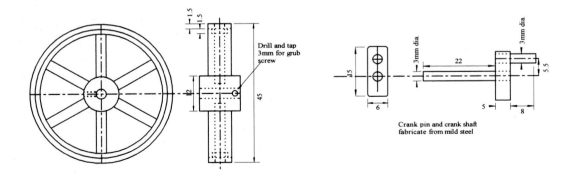

Drill and tap
3mm for grub
screw

Crank pin and crank shaft
fabricate from mild steel

Flywheel - mild steel. Bore 3mm dia for crank shaft.
Six spokes 3mm dia. Fabricate by placing spokes through rim
and into hub. Secure with retaining compound or solder

Metric measurement drawings for inverted vertical engine (2).

8 Douglas, a Double-Acting Engine

So far we have only dealt with single-acting oscillating engines, possibly the simplest method of providing steam power there is. Steam, of course, only pushes the piston in one direction, and we rely on the flywheel to return it to the top of the stroke in order to start the sequence all over again. It did not take the engineers working on the development of the first steam engines very long to realize that the ideal situation would be if steam could also be used to push the piston back, instead of relying on the momentum of the flywheel. The answer was simple enough: seal off the other end of the cylinder and allow steam into there when the piston reached the bottom of the stroke, and the job is done – and that is exactly what we are going to do with this model.

What we need to do is to put in an extra pair of ports at the opposite end of the cylinder and connect them to the steam and exhaust, then

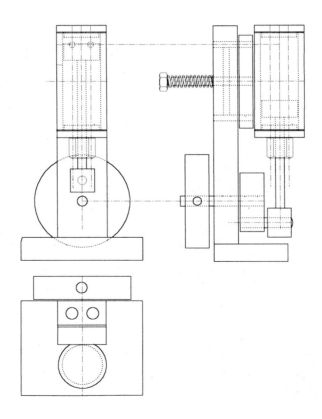

General arrangement: Douglas the double-acting engine.

75

put a cap on that end that is made in such a way that the piston rod will pass through it, while not allowing steam to escape. The initial construction takes the usual form of machining a cylinder to size and fixing it to the port face, except that as we are allowing steam in at each end, they will need to be a little longer and the pivot must be exactly central between the two sets of ports. The work can be carried out in the usual way by marking out with dividers and centre punching for the holes, making sure that the two steam ports are evenly spaced either side of the pivot.

If silver soldering is a practical proposition, then in this case it would be wise to use that method to fix the port face to the cylinder. Both the bottom cover as well as the top one are to be soldered in place later on, and there is a distinct danger of the cylinder coming away from the port face when this is done, if soft solder has been used. If silver soldering is not possible, another method would be to use a lead-free solder for the port face, and one of the lower melting point allows such as 'Tinmans' Solder' for the covers. Do not at this stage solder the end cover on the cylinder, as this needs to be done later.

The port face for the cylinder is made in exactly the same way as for a single-acting engine, except that there is an additional hole for the extra port. While the port face on single-acting engines can be shorter than the cylinder barrel, the double-acting engine requires a full-length port face.

The jig for drilling the ports on the frame or port block also needs to have an extra hole, evenly spaced on either side of the port, and it can be used in the normal way, remembering that instead of drilling one port at each setting, it is necessary to drill two. When all the pieces have been made, including crank, piston and piston rod, have a trial run, turning the crank by hand and looking to see the position of the piston at the end of each stroke: with steam going in each end, it is essential that each port is clear when the piston is at the top or bottom of the stroke.

The bottom cover differs from those dealt with previously in that whereas the latter are simply caps to seal off the end of the cylinder, this one has a hole for the piston rod and a length of external thread for a gland nut. So that the thread and nut are not out of proportion with the engine, the diameter of the piston rod has been reduced, allowing both thread and nut to be reduced in size. This also means that the packing to be used to prevent the escape of steam, in this case an 'O' ring, can be thinner than it would have been if a thicker piston rod had been used, and again reduces the overall dimensions of the bottom cover. If the usual size of rod had been used, the additional length required by a suitable gland nut would have meant that at the top of the stroke the big end would almost certainly have fouled the nut.

Gland nuts are specially shaped to hold the 'O' ring, or other types of packing sometimes used around the piston rod. The essential thing in making the nut is to ensure that the bottom of the inside is flat; the order of machining is therefore to drill with the clearance drill for the piston rod, drill the tapping size, and then use a 'D' bit to flatten the base of that hole: the nut can then be tapped. The tap used to make the thread will need to be ground flat at the end so that it will make the thread right to the end that has been flattened with the 'D' bit. It will probably not be possible to get the thread right to the very end, but it must go as close as possible otherwise it will become necessary to extend the nut, and this starts something of a

The port block also has an extra pair of holes to match the port face; they are best made with a jig, as with two rows the chances of an error are doubled.

vicious circle as it will also mean extending the thread on the bottom cylinder cover.

All the other parts other than the frame are made in the same fashion as before, but don't start assembling the cylinder and fitting the covers as soon as they are made because we will first need to check that the clearances have worked out correctly before that can be done.

No doubt it will have been noticed that the frame is of a slightly thicker brass strip than has previously been used, and we will shortly see why. Anyway start the construction of the frame in the usual way, except that there will be an additional pair of ports; they should be drilled to about half the depth of the block – and then comes the really awkward part. Each pair of ports has to be connected so that steam can pass between them, and to do this a drill is passed from the top of the frame so that it breaks into each one. The obvious way to do this would seem to be to line the drill up exactly with the ports and drill through, so that the drill goes centrally through each one. However, this is not practical, as doing so will not leave sufficient room to get a decent steam connection. They are therefore drilled ½" (or 1mm) towards the outer edge, so that the drill breaks into the ports at the side. The top section of the connecting holes can then be opened out sufficiently to allow them to be tapped to accept a fitting.

Make sure that all burrs are cleaned out by poking around with a piece of wire; invariably the small drills are going to leave pieces of metal around the edges where they break through, and if these are not removed they will block the steam ways and stop the engine from working properly. If a compressor is available, blow them out with that, if not even an ordinary cycle pump can usually do the job.

Now is the time to assemble the piston to the piston rod and bearing, and fit the pivot pin, and then assemble them together with the crank on the frame, but without the cylinder covers. By turning the crank by hand it will be possible to make sure that the ports are being uncovered properly as the piston moves up and down, and if they are not, to make any necessary adjustments. Because of the double action, the piston is smaller than usual, and in order to get the seam between it and the cylinder wall as tight as possible, it has been designed to maximum possible length. A very slight marking-out error could therefore cause it to foul the ports a little, something that we do not want. Checking without the covers will ensure there are no such problems.

When satisfied that the clearances are correct, solder the top cover in place, remembering that if the port face has been soft soldered, there is a possibility of the solder melting, so bind them tightly together with iron wire before commencing. Before soldering the bottom cover in place it will be necessary to assemble the piston and rod and put them in the cover. It will also be necessary to go very

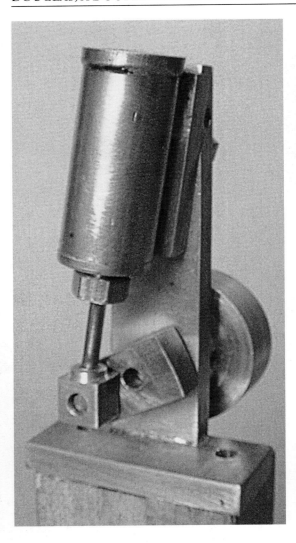

A view of the double-acting engine in which the lower cylinder cover can be clearly seen.

The opposite side of the engine shows little difference from the single-acting one, except for the extra length of the port face.

carefully when heating the job so that the other solder will not melt. One way of doing this is to put the assembly in a tray of water, so that the part to be soldered remains above the water level; the cold water will keep the other solder from melting. This does require the application of extra heat, and it will be necessary to flush the water out afterwards by moving the piston up and down a few times, so that it is ejected via the holes for the steam connection.

The double-acting engine can be reversed in the same way as the other models, and it gives greater power so is more suitable for such operations as powering a boat. It will also start more easily and run more smoothly, particularly at slow speeds.

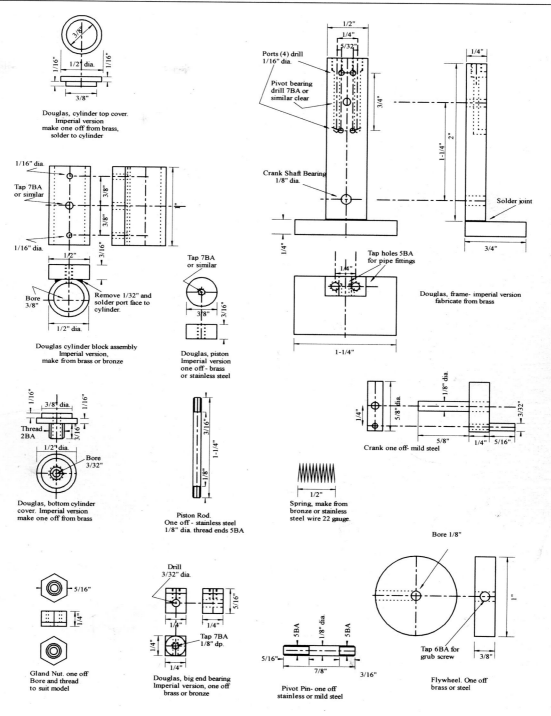

Douglas, cylinder top cover.
Imperial version
make one off from brass,
solder to cylinder

Douglas cylinder block assembly
Imperial version,
make from brass or bronze

Douglas, piston
Imperial version
one off - brass
or stainless steel

Douglas, bottom cylinder
cover. Imperial version
make one off from brass

Piston Rod.
One off - stainless steel
1/8" dia. thread ends 5BA

Gland Nut. one off
Bore and thread
to suit model

Douglas, big end bearing
Imperial version, one off
brass or bronze

Pivot Pin- one off
stainless or mild steel

Douglas, frame- imperial version
fabricate from brass

Crank one off- mild steel

Spring, make from
bronze or stainless
steel wire 22 gauge.

Flywheel. One off
brass or steel

Douglas the double-acting engine: constructional details – imperial.

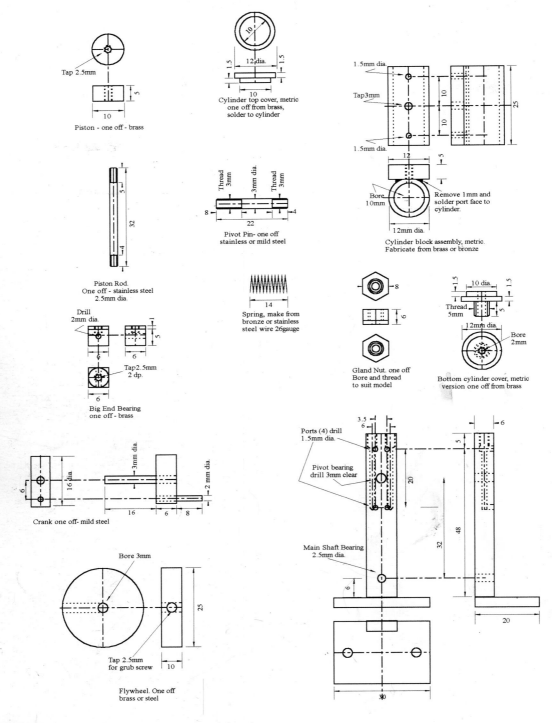

Douglas the double acting engine: metric details.

9 Twinky, a Twin-Cylinder Engine

Two-cylinder engines have the advantage of smooth running and easy starting, and are very popular with boating enthusiasts. To build a two-cylinder engine usually requires the manufacture of a crankshaft, and that is something that many people find very difficult with a small engine. The problem is trying to line everything up, and having got all the parts in line, fixing them together so they will not move out of line once the engine starts running.

This little two-cylinder engine avoids the necessity for a double crankshaft and is no more difficult to build than an ordinary single-cylinder engine – except, of course, that most parts require to be made twice. The design shown is for a single-acting version, but anyone wishing to do so can make it as a double-acting

A general view of the engine, showing in particular the arrangement for fitting the flywheel.

engine, thus gaining more power and even better running qualities, by simply adapting the instructions already given for making the single-cylinder, double-acting model.

MAKING THE PORT BLOCKS

Once again a start is made by making the port or steam blocks; there are some minor differences in sizes, as we have to connect both port blocks together. Care must be taken to ensure that both of them are identical, otherwise the engine will not work. The ports in the steam block pass right through, but the side opposite that, where the cylinder port face will mate, is drilled to allow tubes that connect the two blocks together to be soldered in place. Take care when doing this operation as drills

have a nasty habit of snatching during this sort of job, and if the drill does snatch, the block will be rendered useless. The best way to prevent this happening is to use a slightly blunt drill, as it is more prone to happen with a sharp one. Rubbing a small oilstone over it easily blunts the cutting edge; afterwards the same stone can be used to restore the edge. An alternative is to start the hole with a centre drill and then lay a tiny piece of well worn, fine emery cloth between the drill and the work; this invariably prevents such things happening.

It can be seen from the drawings that the holes meet the centre of the larger hole at the edge of the port: this will allow the drill to break through and provide a steam passage. It is debatable whether it is best to drill the larger diameter first or not; some people will prefer

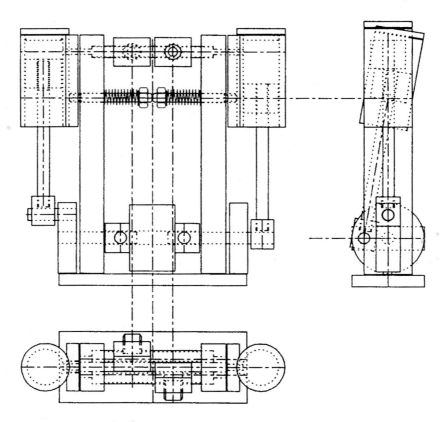

General arrangement of twin cylinder engine.

This view shows clearly the crank and big end arrangement; it also shows how, with a lengthening of the cylinder port face, the engine could be made double-acting.

one way, and some the other. It is very much a matter of personal choice; the only thing that can be said is, if the small hole is drilled last and it snatches, it is easier for the operator to resist the snatch. Of course it is also more likely that the smaller drill will break in the circumstances, so it is easy to see why it must be up to the individual to make up their own mind how to go about it.

The two steam blocks are soft soldered to the base, taking care to ensure that they are the correct distance apart and at 90 degrees to the base edge. The completed cylinders are mounted facing outwards, and are held in place with the usual pivot spring, because the pivots are

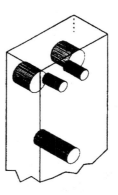

Scrap view showing layout of ports and pipe connections.

shorter than usual, in order to get the necessary clearance; a slightly stronger spring than those used on the single-cylinder engines is suggested. The pistons and piston rods also follow a similar pattern to other engines, and although separate rods and bearings are shown, the idea of a single unit could just as easily be used by anyone with limited equipment.

CRANKSHAFTS AND FLYWHEEL

A normal type of crank is used, but the crankshaft is longer than on the single-cylinder engines because an extended flywheel joins the two crankshafts together. Details of the flywheel are shown in the drawings, and two grub screws are used to secure it to the crankshafts; the cranks should be set at 90 degrees. As shown, the crank webs have been extended to improve the appearance of the engine, however this might create a slight problem with clearance over the base, as the tiniest error will cause the web to foul. It is not much of a problem, and if anyone finds this happening it is simply a matter of removing a little from the length of the web with a file, or even rounding off the corners might do the trick.

The two steam blocks are joined with short lengths of copper tube, and in order to allow steam in and exhaust out, fittings to which other pipes are screwed are also soldered on. It will be noted that these are staggered in order to obtain clearance and to keep the pipes straight. In the case of the engine in the photograph, the effect of having the fittings back to back was not fully calculated before it was made, and it can be seen that it became necessary to kink the pipes to get the necessary clearance, something that is not desirable. Some builders may prefer to set the screw connections at a different angle in order to suit a particular purpose.

The engine can be connected to a boiler via a reversing valve, as it is capable of running in either direction, and if it is to be used to power other models, grooves should be made in the flywheel to accept a belt. To power a boat, one of the crank pins can be extended to create a dog clutch, with the fitting attached to the propeller shaft. The drawings do not show details of the holes in the base that will be required to screw the engine in place, as it will depend entirely on individual usage as to where the placement of these should be; the size of the holes is also a matter of personal choice.

This view shows the connection between cylinders; the engine has not been made in accordance with the drawings, and as a result the two steam connections have fouled each other, which meant the pipes had to be shaped to accommodate them. This will not happen if the drawings are followed.

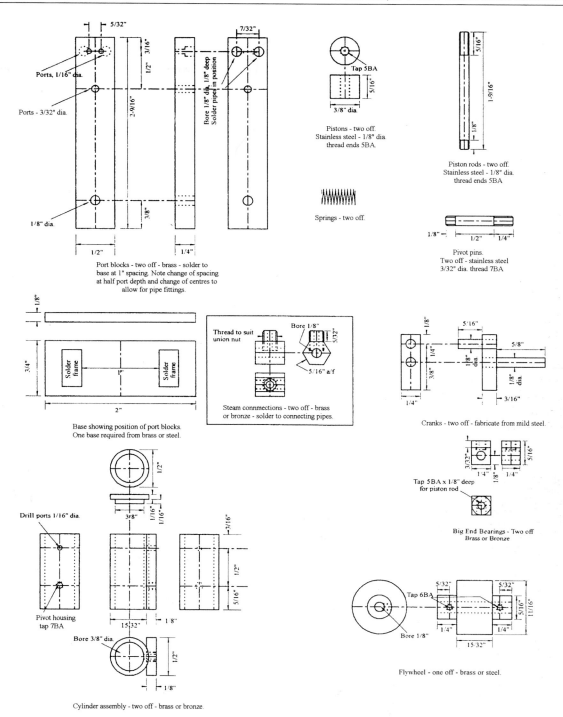

Imperial measurement constructional drawings of twin cylinder engine.

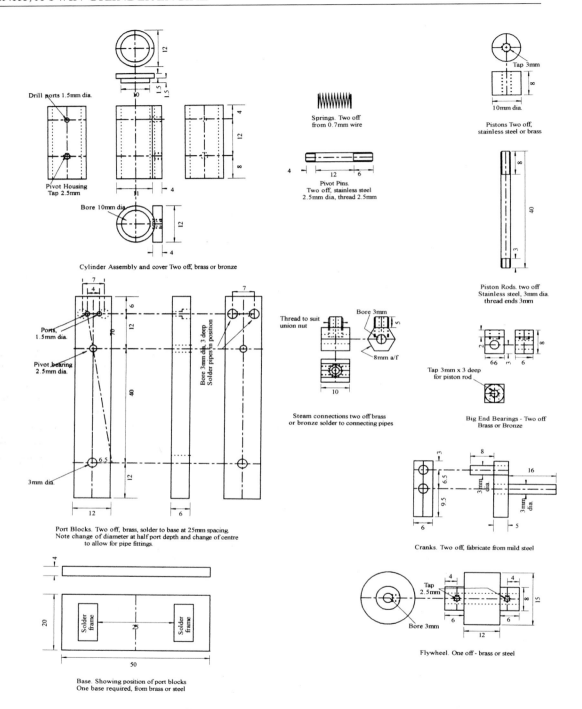

Drill ports 1.5mm dia.

Pivot Housing
Tap 2.5mm

Bore 10mm dia.

Cylinder Assembly and cover Two off, brass or bronze

Springs. Two off
from 0.7mm wire

Pivot Pins.
Two off, stainless steel
2.5mm dia, thread 2.5mm

Tap 3mm

Pistons Two off,
stainless steel or brass

Piston Rods. two off
Stainless steel, 3mm dia.
thread ends 3mm

Ports,
1.5mm dia.

Pivot bearing
2.5mm dia.

Bore 3mm dia. 3 deep
Solder pipes in position

3mm dia.

Port Blocks. Two off, brass, solder to base at 25mm spacing.
Note change of diameter at half port depth and change of centre
to allow for pipe fittings.

Thread to suit
union nut

Bore 3mm

8mm a/f

Steam connections two off brass
or bronze solder to connecting pipes

Tap 3mm x 3 deep
for piston rod

Big End Bearings - Two off
Brass or Bronze

Cranks. Two off, fabricate from mild steel

Base. Showing position of port blocks
One base required, from brass or steel

Solder
frame

Solder
frame

Tap
2.5mm

Bore 3mm

Flywheel. One off - brass or steel

Metric constructional drawings for twin cylinder engine.

10 Victor, the Vee-Twin

Although this model is only a single-acting engine, it is more difficult than the models described previously; in fact it is not an engine for the faint-hearted to build, even though many of the component parts are similar to the previously described single-acting models. The vee-twin is quite popular with model boat enthusiasts as they find it gives an extra bit of power, while at the same time not requiring a great deal of room. In full-size practice the design was quite commonly used in the open launch type of boat, although with a slide valve configuration rather than an oscillating one. The engine produces considerable power when built as a single-acting one, and the more enterprising model maker will no doubt be able to build it as a double-acting engine if they wish to. It is possible to extend the cranks' shaft

and fit a pulley next to the flywheel should the engine be required to power other machinery. To drive a model boat the flywheel can be converted to a dog clutch by inserting two pegs about $3/32$" or 2.5mm diameter in the flywheel. It is capable of being reversed by using a valve to reverse the steam flow.

THE BODY

The two port faces form the body of the engine, and to start with they should be made identically in the normal fashion, except that before the steam ports can be drilled, using the usual type of jig, it is necessary to join the two pieces together. The model shown in the drawings has been made by cutting and filing out two short sections and soldering them

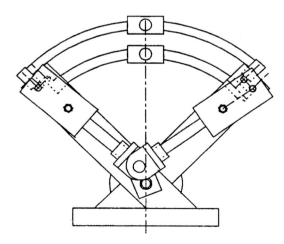

General arrangement drawing of vee-configuration engine showing big end arrangement.

together with soft solder. They are at 90 degrees to each other, and any shallower angle will result in the big ends fouling each other, but it is possible to make an engine with a greater angle, and even with the cylinders horizontally opposed if one wishes.

The easiest way of marking the pieces out for cutting is to put a length of rod through the two crankshaft holes, set them against a square, and scribe along the overlapping sections. Run a hacksaw along the lines, and then file half way through each piece so that they interlock. When the two sections are flush with each other, tin them both individually by running solder in each recess and wiping it off with a cloth before it sets. The usual precautions against using a cloth with man-made fibres apply when doing this, and ideally pure cotton material should be used.

Rub a pencil all round the circumference of a piece of rod to be fitted in the bearing holes, in order to stop the solder from adhering to it. Flux the two tinned surfaces, and heat the whole assembly with a small blowlamp; the existing solder from the original tinning operations should be sufficient to secure the parts permanently together. Of course, anyone with milling facilities can mill the two recesses, saving a great deal of work and also no doubt making a neater job. Whichever method of cutting out is used, it is essential that when soldered together the two sections are perfectly in line with each other and will lie absolutely flat when put on a known flat surface.

The two ends of the sections that have been joined need to be trimmed off before the base can be soldered in place; ideally this should also happen before the steam ports have been drilled. To drill the ports, use one of usual jigs and make certain that they are only drilled to half the depth of the arm, not right through. Having done that, life now starts to get very difficult, as it is necessary to run pipes between the two arms of the vee and connect them to the ports. If you are happy to allow steam to escape from the ports, only one connection will be required; although then, of course, it will

not be possible to reverse the engine. Two connections not only make a neater and more professional-looking job, but they also mean that the engine will be able to go backwards.

Unfortunately the outer port on each has to be connected to the inner one of the other, which involves quite a bit of drilling and blanking off – though fortunately actually doing it is nowhere near as complicated as it sounds.

MAKING THE PORTS

With the face that will hold the cylinders towards you, start work on the left-hand arm and the port on the left-hand side of it. This one is quite straightforward: simply drill down from the top at the outside of the port, so that the hole will just break into the port. A small brass fitting, shown on the drawings, is used to hold the pipe, and so the hole should be made the correct size for that to be pushed home and soldered in place. Now we can turn our attention to the inside port of the other arm and repeat the operation.

Keep on the right arm in order to work on the second port and start in the same way, but this time use a drill that is sized for a suitable tap that you have in stock. That, too, breaks into the edge of the port, but on this occasion the hole is continued past it. To a depth lower than the first port on the arm, having ascertained a suitable depth, drill from the inside of the arm so that the drill breaks into the hole that has just been made; the hole

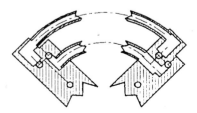

Scrap sectioned view showing layout of steam passages and connections.

An angled view of the engine that shows the piping arrangement for the steam connections.

should be of the same diameter as the pipe to be used so that it can be soldered directly in place.

The connection for the pipe in the left arm is made in exactly the same way, except that in this case, of course, the hole does not pass under a port; even so, in order to keep things tidy the position of the pipe should be the same on each arm. We now have a way of connecting both pairs of ports, but there are some holes through which steam will escape and so these have to be blocked off. They are, of course, the two top holes that were last made, hence the advice to make them a tapping size as both can now be tapped and small plugs inserted and

filed flush with the tops of the arms. The alternative is to solder plugs into them, if this is found preferable.

The lengths of pipe used for connections are split in the centre and soldered into little brass blocks that can be used to connect the engine to the steam from the boiler, the other going to exhaust. The base can at this stage be soldered in place; the holes for bolting the engine down should be drilled prior to doing so.

THE BIG ENDS

The big ends – or perhaps we should say 'big end', as two are combined in one – must

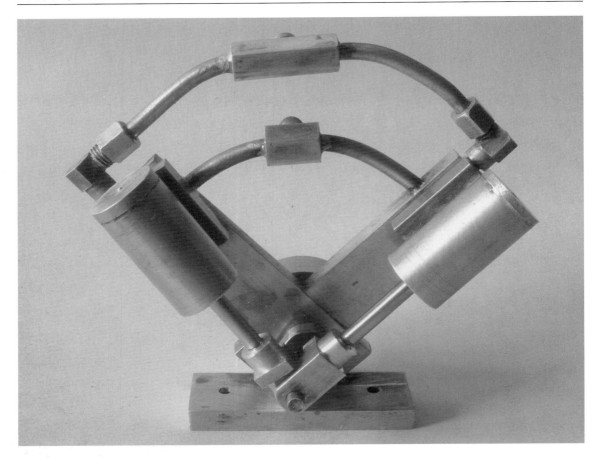

In this photograph the arrangement of the big end bearing is clearly visible.

obviously differ from the normal pattern because of the two cylinders. The drawings show two versions, and it is left to the builder to decide which will suit him or her. In one case the sections that slip over the crank pins are simply filed to half their normal width, while on the more sophisticated version a slot is cut in one of the brasses and the other filed to a tongue to fit in the slot. Both pairs have rounded ends for clearance purposes, and anyone with milling facilities wishing to make a more authentic-looking version can make a rounded recess for the ends to fit into, rather than leave them plain as shown.

FLYWHEEL

A two-cylinder engine of any sort with the cranks at 90 degrees will operate without a flywheel, but the addition of one will improve the operation. This means that there is no need for anything massive, and the one drawn is fairly small and narrow; as usual it is held in place with a grub screw tightened on the crankshaft. As with all the engines, a shim washer should be placed between the crank and the port block, as well as a similar one between the flywheel and block, in order to avoid the corners catching and making score marks.

AN ALTERNATIVE ARRANGEMENT

For those who do not feel they can cut the sections from the port blocks and get them to line up successfully, there is an alternative method of construction. The two port faces are soldered at 90 degrees, but instead of a joint they are placed one on top of the other. In the same way as described above, a length of steel is used to line up the crankshaft holes while the soldering takes place. This will mean that when the two cylinder assemblies are fitted, the piston rods are on a different line to each other. The crank pin has to be extended to accept two big ends side by side, and as they are not going to intermesh they can be made slightly narrower.

The situation now is that, as drawn, the cylinders will not line up with the positions of the big ends. Therefore it is necessary for the cylinders to have two different widths of port faces soldered to them, one thicker than the other, which will allow them both to line up with one of the big ends. The big ends will then run together side by side, instead of being intermeshed, and while it is a clumsier arrangement than the previous one, it will enable a vee engine to be built without having to make the joint and get the two pieces to fit exactly together, which is the most difficult part of the job.

The only other additional work needed is to ensure that the pipes connecting the cylinders are bent in order to fit, or to make special fittings that will allow them to line up accurately.

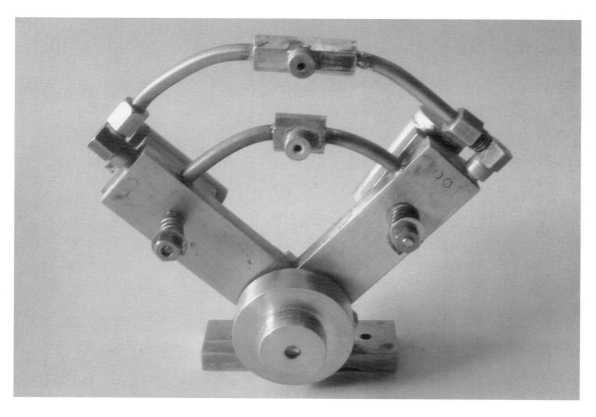

A view of the vee engine, showing the two pivots and the flywheel.

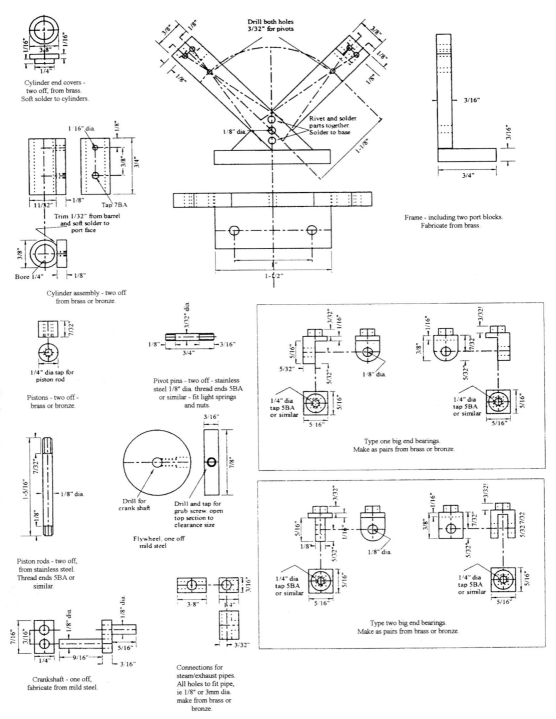

Cylinder end covers -
two off, from brass.
Soft solder to cylinders.

Drill both holes
3/32" for pivots

Rivet and solder
parts together.
Solder to base

Frame - including two port blocks.
Fabricate from brass.

1 16" dia.

Tap 7BA

Trim 1/32" from barrel
and soft solder to
port face

Bore 1/4"

Cylinder assembly - two off
from brass or bronze.

1/4" dia tap for
piston rod

Pistons - two off -
brass or bronze.

Pivot pins - two off - stainless
steel 1/8" dia. thread ends 5BA
or similar - fit light springs
and nuts.

1/4" dia
tap 5BA
or similar

1/4" dia
tap 5BA
or similar

Type one big end bearings.
Make as pairs from brass or bronze.

Drill for
crank shaft

Drill and tap for
grub screw. open
top section to
clearance size

Flywheel, one off
mild steel

1/4" dia
tap 5BA
or similar

1/4" dia
tap 5BA
or similar

Type two big end bearings.
Make as pairs from brass or bronze.

Piston rods - two off,
from stainless steel.
Thread ends 5BA or
similar.

Crankshaft - one off,
fabricate from mild steel.

Connections for
steam/exhaust pipes.
All holes to fit pipe,
ie 1/8" or 3mm dia.
make from brass or
bronze.

Constructional drawings for vee engine with imperial measurements.

92

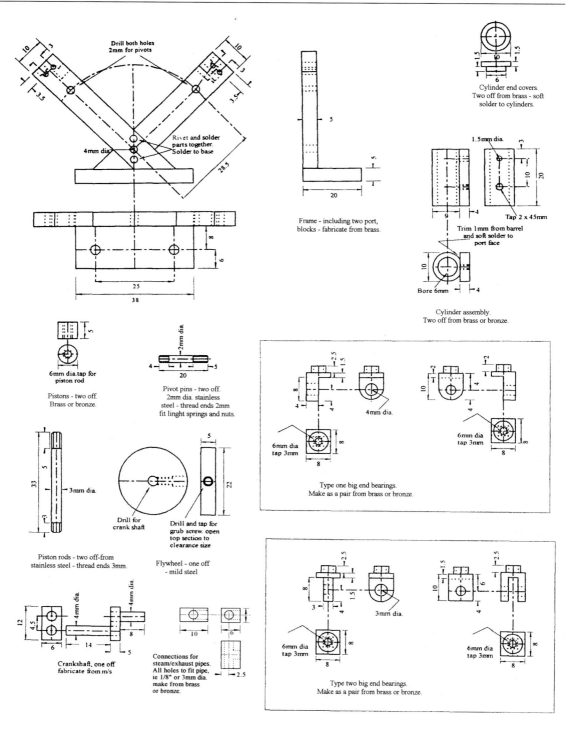

Drill both holes
2mm for pivots

Rivet and solder
parts together.
Solder to base

4mm dia

Cylinder end covers.
Two off from brass - soft
solder to cylinders.

1.5mm dia

Tap 2 x 45mm

Trim 1mm from barrel
and soft solder to
port face

Bore 6mm

Frame - including two port,
blocks - fabricate from brass.

Cylinder assembly.
Two off from brass or bronze.

6mm dia.tap for
piston rod

Pistons - two off.
Brass or bronze.

2mm dia.

Pivot pins - two off.
2mm dia. stainless
steel - thread ends 2mm
fit linght springs and nuts.

3mm dia.

Drill for
crank shaft

Drill and tap for
grub screw. open
top section to
clearance size

Piston rods - two off-from
stainless steel - thread ends 3mm.

Flywheel - one off
- mild steel

4mm dia.

6mm dia
tap 3mm

4mm dia.

6mm dia
tap 3mm

Type one big end bearings.
Make as a pair from brass or bronze.

Crankshaft, one off
fabricate from m/s

Connections for
steam/exhaust pipes.
All holes to fit pipe,
ie 1/8" or 3mm dia.
make from brass
or bronze.

6mm dia
tap 3mm

3mm dia.

6mm dia
tap 3mm

6mm dia
tap 3mm

Type two big end bearings.
Make as a pair from brass or bronze.

Constructional drawings for vee engine with metric measurements.

11 Clarence, a Clapper Engine

Entirely different to the oscillating engine, the so-called clapper engine is just as easy to build and is great fun to run. The engine uses a clapper valve to admit steam, and it is from that valve that the name is applied to the engine; it is unlikely that there has ever been a full technical name for the valve, and it is called a clapper because of the noise it makes when running. The method of operation is extremely simple: the piston has a projection on the end, and when it reaches the top of its stroke, the projection opens a valve and allows steam into the cylinder; this moves the piston back, and

the spent steam is released through holes in the cylinder wall. On the model, in the interests of simplicity, a ball bearing has been used as a valve, which keeps construction as simple as possible.

THE CYLINDER

For the cylinder a short length of square brass bar has been used, but there is nothing that says a square bar is essential, and the cylinder could just as easily be a length of round rod, or hexagon bar. The material is bored in the usual

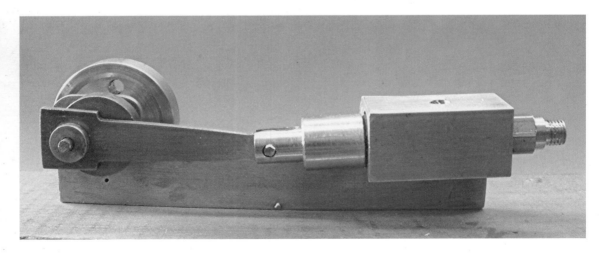

This view shows clearly the crosshead arrangement, as well as the steam exhaust hole in the cylinder. If the model is scaled to a larger size it would be wise to use several such holes of around the same size, rather than increase the diameter. Readers will also no doubt notice that a round crank has been used instead of the bar type in the drawings; again it shows the flexibility available to the builder.

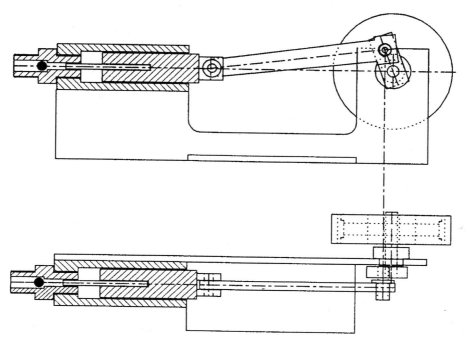

General arrangement drawings, sectioned to show clapper valve.

fashion, but the bore does not go right through, it is finished blind by use of a 'D' bit. In the blind end is a tapped hole, and to keep this in line with the policy of using similar threads throughout the series, it is shown as 2BA or 5mm.

A short extension screws into the end of the cylinder; this is shown as made from hexagon rod, but it could easily be made from round stock. Both ends are threaded, and a hole is drilled in the end that fits into the cylinder and meets another hole, the base of which is flattened with the 'D' bit. A $\frac{3}{32}$"- (2.5mm-) diameter ball is placed on the flattened hole and given one smart tap with a hammer, via a short length of rod, to provide a seat. The ball used to make the seat should not be used for the valve in case the hammer blow has damaged it, therefore a new rustless ball should be fitted. There are three choices here: bronze, stainless steel or nitride rubber; all work quite well.

THE PISTON

The piston is made from bar stock, either brass or stainless steel. In one end is a small hole into which a piece of brass rod is fitted, and the material should be a good fit in the hole; it is not secured in any way because after a period of time it will wear and need to be replaced. It should be long enough for about $\frac{1}{32}$" (1mm) to protrude through into the chamber where the ball will be situated when the piston is in its furthest forward position.

CROSSHEADS

The opposite end of the piston has a cross-drilled projection with a slot, the latter being to accept the piston rod; it is technically known as a crosshead (which sounds something like the morning after the night before). Unlike the oscillating engines where the cylinder moves in

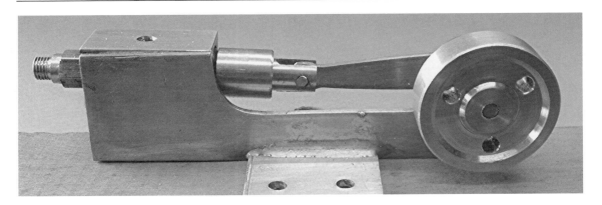

The opposite side of the engine has the plate with which to screw it down, and also the flywheel can be seen. Yet again this differs from the drawings, but only inasmuch as it has three holes drilled at equal spacing. These have absolutely no effect on the running of the engine: they are only for embellishment, and in fact were an afterthought when the model was built.

line with the piston rod and crank, in this case the cylinder will remain stationary and a connecting rod is used; there is no piston rod as such, which anyway is, of course, only an extension of the piston. In more advanced engines the crosshead is situated at the end of the piston rod and is jointed in order to get the necessary movement. In our case the piston rod is dispensed with, and the crosshead is part of the piston. This not only ensures it is easier to make, but also the extended length of the piston has a couple of advantages. Firstly, it gives it a better chance of being steam tight; and secondly, as there is less flexibility, it will give it a longer working life. There is nothing complicated about all this, it is simply a slot with a hole across, through which passes a pin that also goes through the small bearing in the connecting rod.

THE PIN

The pin is a matter of personal choice: it can be machined with a larger head than the shank and threaded for a nut, a rather fiddly thing to do in the sizes shown on the drawing, but perfectly practical if anyone has decided to scale

everything up. Alternatively it could be a plain pin held in place with a drop of retaining compound on one side of the crosshead. In the latter case, care must be taken to avoid the retainer spreading to the pin and locking everything up altogether. Yet another alternative is to apply a touch of soft solder to the pin and crosshead, again ensuring that the solder does not run into the connecting rod bearing. Perhaps the best method is one of really bad engineering practice, but perfectly suitable to this situation, and that is, to make a pin that is a good fit and just give one end a very smart tap with a small hammer. That will spread the metal and allow the pin to be pushed home and remain tight in the crosshead. It can also be easily driven out if it is found necessary to do so.

THE FRAME

The cylinder is soft soldered to a frame made from sheet material; it can be either steel or brass. At the opposite end to the cylinder is the crankshaft bearing. As the frame is of thin material it will be necessary to incorporate a bearing, otherwise in no time at all a hole used for such in the frame would quickly elongate and cause problems. The bearing must be made of brass or bronze, and can be either soft soldered to the frame, or a nut can be used in a similar manner to the bearing on the mill engine.

The centre of the cylinder bore should line up with the centre of the bearing and, as can be seen in the photographs, adjusting the angle of the cylinder made a correct alignment. Whether or not to work it out in this fashion will be an individual choice; some readers might prefer the cylinder in line and horizontal.

THE CRANK

Made in exactly the same fashion as all the other cranks, manufacture will need no description and either a disk or bar-type of web can be used. The flywheel, too, is made in the same form as for most of the other engines and will be familiar to readers.

THE CONNECTING ROD

In the case of the connecting rod there is a difference, because it is not to be screwed into the piston to work in line with it: it has to be drilled and shaped from flat material, and either brass or steel will do. A small bearing has to be made to fit the big end, for the same reason that one was used in the frame, and that must be made from brass or bronze. A hole acts as the bearing in the small end, and an already pointed-out pin is pushed through from the hole in the slot in the piston, as a means of securing it, while ensuring it is flexible. If steel has been used for the piston rod, the pin must be brass or bronze; if brass has been used, then the pin should be steel.

THE FLYWHEEL

The flywheel follows the regular pattern, and because it is of a rather large diameter, the centre portion has been thinned to save weight. The engine in the photographs has had three holes drilled in the wheel, but these were done purely for cosmetic reasons and have no effect on the running qualities. The drawings do not show a method of fixing. The photographed model has had a reamed hole to a push fit on the crankshaft, but it is appreciated that not everyone will have a suitable reamer and so the fixing is left to individual choice. The best way will probably be to drill and tap a hole for a grub screw in the centre boss.

OPERATION

Running the engine is quite straightforward: a steam pipe is connected to the cylinder extension, and the flywheel given a quick turn by hand to set the engine running. Direction of motion will depend on which way it is turned, and although the engine will run in both directions, it will not reverse mechanically. It is designed to run at very high speeds and therefore could be used in a model speedboat if one so wished. Sizes can easily be altered if a larger engine is required. Should alterations be made in the size, the lift of the valve should not be increased but should remain the same. Obviously in the case of a larger engine the valve diameter must also be changed, but it is not wise to leave too much space around the sides of the ball, for seating will become difficult and the ball will tend to bounce rather than seal the inlet. This is also the reason why the amount of lift should not be changed because increasing it will unduly delay the valve operation and impair the operation of the engine.

A steam pipe is connected to the end, and when steam is turned on it will drive the ball into the recess and prevent it from entering the cylinder, and will be released into it when the ball is pushed from its seat by the protrusion from the cylinder. Some people may wish to cross-drill the connection and insert a pin to prevent the ball from coming out when the engine is not connected to a steam pipe. It will no doubt largely depend on whether or not the connection is to be permanent.

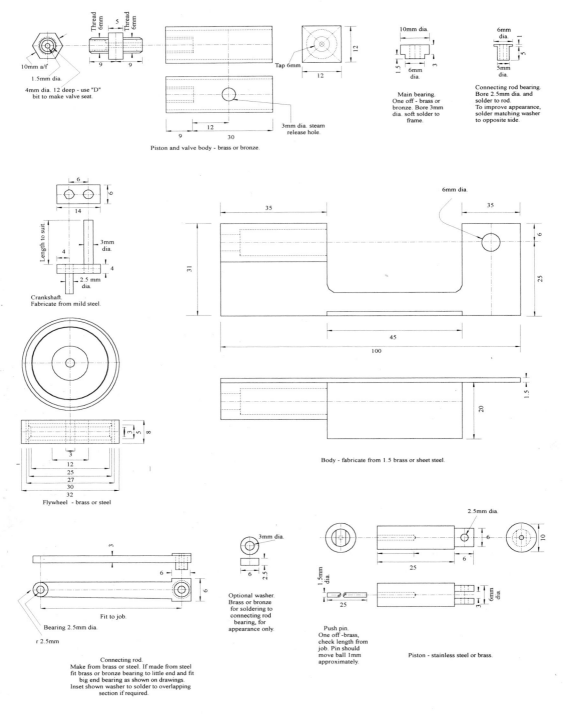

Constructional drawings for clapper engine, metric measurements.

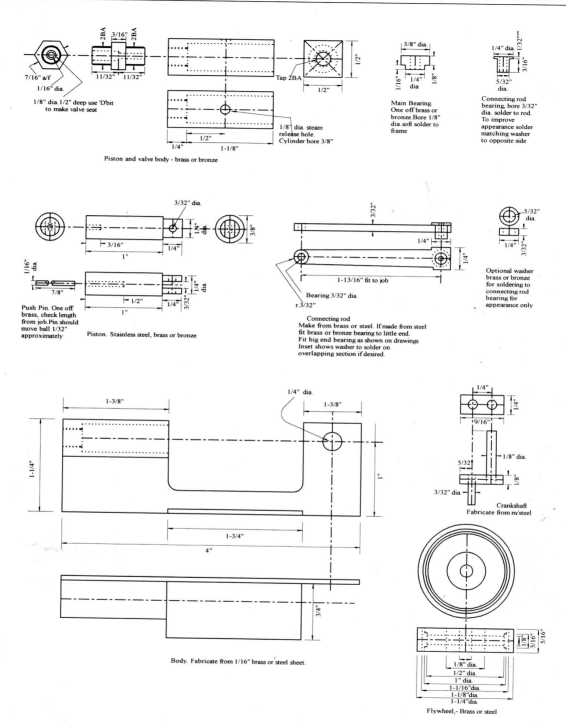

Constructional drawings for clapper engine, imperial measurements.

PART THREE
BOILERS

12 Basic Boiler Construction

As the models described herein are intended to be driven by steam, it follows that we need a means of supplying it, and many readers might well find this the hardest part of the construction. Fortunately all the boilers described are comparatively simple, and although they do increase very slightly in complexity as the engines themselves become more complicated, it will be possible to operate any of the models from the simple type of boiler called a 'pot boiler', although the more advanced boilers will ensure a better performance.

Steam is a common gas that we see every day, although because it *is* such a common thing we might not realize that in fact it is a gas. It is the result of heating a liquid, in our case water, and although commonly believed to be the white vapour we see coming from a kettle or saucepan, it is in fact a colourless gas, and if it could be seen would be found in the tiny gap between the spout of the kettle and that white vapour. The vapour is actually the gas condensing, and is at a considerably lower temperature than we need to run a steam engine. It condenses as it mixes with the colder

air, but while it is actually condensing it is also expanding. The amount of expansion in steam is quite incredible; thus if we were to leave the room where the kettle was being heated, and return when it had boiled dry, the room – unless it were a particularly large one – would be found to be full of vapour. This means that a litre or so of water when heated has become several cubic yards of condensing gas. Just think what an impossible task it would be to try and get it back into the kettle: there is no way it would fit until it had returned to its original temperature.

By confining the gas in a sealed container, namely the boiler, and preventing it from escaping, there is less opportunity for it to condense; and yet it will still be attempting to do so and therefore expanding, with the result that pressure builds up inside the container. It is this pressure, coupled with the expansion ability, that drives the steam engine.

There is nothing magic about a boiler; all we need to make is a vessel that is strong enough to hold the steam at the pressure needed to make the engine work, plus a means of allowing that steam to escape in a controlled fashion into the engine. A simple tap will suit this purpose, though steam-engine enthusiasts prefer to call the tap a regulator, because by opening and closing it slowly, thus allowing a limited amount of steam to escape, the speed of the engine can be regulated. In a motorcar, or any other machine using an internal combustion engine, it would be referred to as the throttle. In fact it matters little what name is given to it, or how the finished article looks: it is still basically nothing more than a tap. It is fair to say, however, that as a rule an effort is made to make a tap that is infinitely more controllable than the type found domestically. But we will come to regulators and how to make them later on.

Because of the ability of steam to build to a very high pressure, boilers that are fitted with a means of stopping the escape of steam are also fitted with a safety valve, which at a given pressure releases a certain amount of steam, thus keeping the pressure to a safe limit. The safety valve can take a number of forms, and in our case will be kept as simple as possible. The engines described here work at a comparatively low pressure; nevertheless, if there is a stop valve, a safety valve should always be fitted and adjusted to open at a suitable pressure.

Most toy steam engines have boilers made of brass, and although there is nothing wrong with the way they work, brass is not a material to be recommended because it can react with hard water: the zinc content of the brass is caused to leach away, leaving the residue in a powdery state that, when mixed with water, becomes paste-like and eventually porous. The boiler is not likely to explode, as the effect will be localized and the initial failure will be in the form of a leak, but it is not something we wish to happen, and when making one's own boiler, copper is preferable. Boilers on steam toys are very often assembled with soft solder, and attempts by youngsters to make a model go faster by tightening the safety valve and thus increasing the pressure beyond that which was intended by the manufacturer, more often than not results in the soft solder melting and the

boiler falling apart. Heat increases with pressure, and that is why the solder melts. It is not unsafe to assemble a boiler with soft solder, because if worked too hard it will simply fall apart. The heat source specified for these models is in itself insufficient to melt the soft solder.

SILVER SOLDERING

The model engineer constructing a model will usually silver solder the boiler, and this combined with copper of adequate thickness to give the required strength, results in a good, sound pressure vessel strong enough to withstand much higher pressures than we will require. The engines described here work on pressures of below 30lb per square inch, and therefore do not require such strength. Even so, if a builder has the equipment, materials and knowhow to silver solder a boiler, it is far superior to soft soldering; though lesser mortals can rest assured that soft solder will be perfectly safe to use. Providing no attempt is made to increase the pressure, the method of heating these boilers ensures that they will not be over-pressurized, and only by changing the type of fuel and burner would this be possible.

Perhaps at this stage a word about solder may be prudent. If soft solder is to be used to make a boiler, it does not mean that any old solder can be used; in particular the cored type of solder used for electrical work is not suitable for boiler making. The best material is the lead-free type: lead-free solders have been specially formulated for use on copper and brass, and have a melting point of over 200°C. They are easy to use with a small blowlamp, and are non-toxic to work with, though the use of the correct type of flux is essential to ensure good, sound joints.

Boilers can be very simple affairs, or they can be immensely complicated; those shown in the following chapters are all comparatively simple to build and yet quite adequate for driving the models. The burners to be used with the boilers have also been selected as the most suitable for that particular type of boiler, mainly because of their shape. Provided they can be placed underneath the boiler, any burner will drive any boiler, and should there be a wish to economize on work and material, readers can, if they wish, reduce the number of burners they make, as most likely only one model at a time will be operated.

TOOLS AND MATERIALS FOR BOILER MAKING

In general, only simple hand tools are needed for boiler making. First, something with which to cut the material, which might take the form of either sheet-metal cutters or a coping saw. As the end plates usually need to be flanged over, so a hammer will be needed, and while it is best to use a soft-faced one, it is not strictly necessary and an ordinary metal-headed one will do – although the latter does have the disadvantage of marking the material that is being worked on, rather more than will a soft-

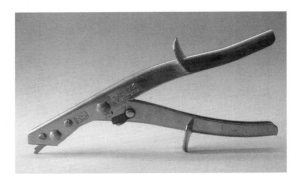

The Nibbler is a useful tool for cutting sheet metal; it works on the principle of a small square punch making a continuous series of holes that slightly overlap, thus creating a continuous slot. Unlike tin snips, the Nibbler does not distort the metal that is being cut. The tool shown in the photograph is hand operated, but it is possible to buy a pair that works from a hand-held power drill, and even one that is fitted with a motor, making a self-contained unit.

faced one. In addition, some means of drilling holes will be needed, plus a file for general trimming up of the materials; apart from that, only the heat source is required.

BLOWLAMPS

Although it might be possible to manage without a blowlamp by heating everything on a stove of one sort or another, one is desirable, to say the very least. It needn't be too big – though beware of some of the very tiny ones that are sold with an advertised heat of about 2,000°C. It is true that the tip of the flame will reach such a temperature, but that is only a pinpoint of heat and does not have sufficient volume to do the job that we require.

This may be hard to understand, but look at it in this way: the heat at the tip of such a tool is certainly at that temperature, but if we wished to boil a saucepan using that flame it would be a sheer waste of time. The tiny quantity of heat would disperse across the bottom of the saucepan, and if we held it there until the year 3,000, the water in the saucepan still would not boil. Do the same thing with a flame that is at a lower temperature but is about ten or more times the size, and our saucepan will heat up and boil; so we therefore need quantity as well as quality.

Metal shears such as these are useful for cutting copper sheet; however, be warned that for most shears, or 'tin snips' as they are often called, the maximum cut is 16SWG or 1.5mm.

BRAZING HEARTH

Because copper disperses heat so quickly, an overall source of heat is preferable to a localized one, and the best way to obtain this is to organize what is known as a brazing hearth. This consists of nothing more than a metal container holding a quantity of heat-retaining material: an old baking dish is ideal for this purpose, one with sides about 2" or 50mm deep. The heat-retaining material can be from a variety of sources; proper silicone bricks as used in potters' kilns are ideal, but may not be readily available, but it is possible to obtain firebricks that reflect heat. Avoid the type used to line the backs of fireplaces, as generally they only do just that and act as liners; they are not designed for heat reflection. Thermolite blocks are available at builders' merchants, and they do give a certain amount of reflected heat and can be used for our purpose.

This DIY type of blowlamp is capable of silver soldering a small boiler, but it will only do so if the material being silver soldered is specially packed with heat-reflecting material, in order to avoid loss of heat to the atmosphere.

FLANGED PLATES

A boiler consists basically of a tube, the ends of which are sealed with flat or sometimes slightly domed plates. In order to obtain sufficient strength at the joints we need to make a flange on the plates: hence the name. To do this, a circle of copper about ½" or 12mm larger in diameter than the internal size of the tube is cut. The edges of this are flanged over so that they slip inside the tube and can then be soldered in position. The flanging is done over a former that can be made from either metal or hardwood. If it is proposed to use the former to make a number of models, then steel is the best material to use, but for a single boiler hardwood or even thick plywood is quite adequate.

In order to be able to bend the material round the edges of the former it needs to be very soft, and we can soften or anneal it by simply heating it until it becomes a bright red colour – in other words, 'red hot' – then quenching it in cold water. The heating needs to be even over the whole of the metal, and

A typical former with which to make flange plates for a small boiler; it is a short length of steel bar, with rounded edges. The metal for the end plates must be annealed to ensure a close fit.

some readers may find that this can best be achieved by laying it on an ordinary domestic gas stove, rather than using a blowlamp that can heat unevenly. A camping stove will also do the job if the domestic authorities object too strongly to the use of the gas stove, or if the cooker in your house happens to be all electric.

Once the copper has been softened, place it in a vice against the former, protecting it with a soft jaw, and then hammer round the edges. Don't try and put one small section right over to its finished size – rather, go right round bending a little at a time, and keep going round until the finished shape has been achieved. Copper gets hard when it is hammered, a process known as 'work hardening', and so it may be necessary to re-soften it three or four times before the finished result can be obtained. Quite an amount of force might be necessary to shape the copper, and the common term for boiler making amongst model engineers is 'boiler bashing', which refers to the need to make flanged plates.

A number of holes will be required in the boiler: one or more to allow the amount of water to be estimated (this or these will be in the flanged plates), a filling hole, as well as one for the pipe that takes the steam to the engine, though in some cases the latter two may be combined. On more sophisticated models there will be a need for holes to accept tubes that are used to make the boiler more efficient, and possibly one or two more for various fittings that are to be attached to the boiler. All these should be made before any attempt is made to solder the boiler together.

How to fit the end plates to the tube is a matter of personal choice; a much neater job is made if the flanges are turned in towards the inside of the tube, though it is far easier to have them facing outwards. Although this may look a little untidy, it makes holding them in place while they are soldered a great deal easier and safer. It is not unknown for the end plates to slide into the boiler during heating: the tube expands and away they go, and trying to get them out and start again is a horribly awkward job.

Flanged plates will always finish up with uneven edges, though fortunately when the plates are inside the shell this matters not, as nobody else will see them. If they are being placed on the outside, however, it is a different matter, and they should be tidied up. A scribing block or surface gauge can be used to mark an even line right round the plate edge, and a hacksaw used to tidy the edges up. Depending on how fussy one is, it may be necessary to finally resort to a file to get the edges looking exactly right.

The best idea, whichever way the plates are to be, is to insert some small copper rivets to ensure it cannot happen. It will not be possible to close the rivets if the flanges are facing inwards, but this matters not, as their job will only be to prevent movement along the tube, not to secure the parts together. If rivets are being fitted, for cosmetic reasons they should be evenly spaced both from the edge and round the circumference. Rivets were always a feature of old full-size boilers, which were only held together by them. There was no solder to caulk the seams; the boilermaker had to keep hammering the edges together until the boiler was steam tight.

After soldering the end plates in position, any tubes that are used to improve efficiency should be fitted and soldered in place, followed by the bushes for any fittings. Between each soldering operation the boiler must be cleaned: if silver solder is being used it should be put in an acid bath; if it has been soft soldered the acid will frequently make the solder turn black, so the assembly should be thoroughly washed in warm water to which a tiny drop of washing-up liquid has been added, and then thoroughly rinsed in clean water. Afterwards the places to be soldered next should be physically cleaned. Don't use emery paper or cloth, as the grains on that are slowly removed as it is used and will stay on the boiler. Use either a wirewool 'Scotchbrite' or a clean brass wire brush.

FIRE AND WATER TUBES

Some of the boilers have fire or water tubes that improve steam-raising properties. These should be a good fit in the holes in the boiler, because if there is too much slop the solder will run away into the inside of the boiler and be wasted. In the case of water tubes that have

ABOVE: *Two disks of copper, ready for converting into boiler end plates. They have been annealed before starting, but further annealing should take place frequently while the work is being carried out.*

RIGHT: *It is unlikely that the plates will have smooth and regular edges: these will need to be created, and here we see a start being made by using a surface gauge to ensure both will be evenly marked the whole way round.*

With care it will be possible to cut along the line using a hacksaw, or as is the case here, with a junior saw.

The finished item, with two nicely finished, flanged boiler end plates.

been shaped, the tubing should be thoroughly annealed before attempting to bend it. Even then there is a danger of it kinking unless one is very careful, and that will restrict the water flow. If a spring that is a good fit to either the outside or inside of the tube is fitted over or in it prior to bending, it will not kink.

If bends are very tight, a spring can be difficult to remove, and in that case it is best to adopt one of the following methods. The first is to completely fill the tube with a low melting-point metal, such as that sold for casting in silicone rubber moulds. Again there can be a problem putting the metal into the tube if it is a very small diameter. The other way, which readers may find more acceptable, is to make sure the tube is well annealed, and solder a piece of metal to cover one end. Fill the tube right to the top with water, and place it in the domestic freezer. When the water has frozen hard it will be possible to bend the tube without it kinking. If bending starts to become difficult, start the process over again. As already mentioned, copper gets hard with work, and so many shaping processes require it to be annealed several times in order to get the job finished.

BUSHES

Bushes are best made from bronze; the effect of brass with hard water has already been mentioned, and this problem can be avoided by using bronze. The bushes should always be flanged to allow a seat inside the boiler that will assist in soldering them in position. After they have been soldered it is advisable to just run the tap through them again to ensure they are clear and that any fitments can be screwed in. Bushes are a problem for anyone attempting to make a boiler, who does not own a lathe, and the only way to overcome it is to solder commercially made brass nuts in place. This goes against the advice to use bronze rather than brass, but it may be a case of necessity. Brass screws can be adapted to make fittings such as the take-off for steam, and of course can also be used as plugs to check water levels.

TO TEST OR NOT TO TEST

If a boiler is to withstand a very high pressure it should be hydraulically tested to twice its working pressure. Members of model engineering and model boating societies may find that the rules of the society insist this is done before the boiler can be used in public. The society will have test equipment and an appointed official to test boilers, and one might as well make use of the facilities offered. What a person does at home is their own business, and there is no law of any sort that insists the boiler be tested.

Why does a boiler need testing? For two reasons: firstly and most importantly to ensure it is absolutely safe; secondly to see if there are

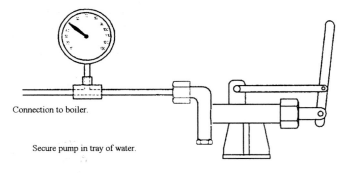

Connection to boiler.

Secure pump in tray of water.

Set-up for boiler testing using one connection only on boiler.

any leaks. Now, these are two entirely different matters altogether. A boiler should be safe, and about that there is no doubt: thus all the boilers described here are quite safe to use, provided that the thickness of the copper stipulated in the drawings is adhered to, and that their working pressures are not exceeded.

Some people will wish to test them for their own peace of mind, and for this a hand pump is needed. A pressure gauge capable of reading at least twice the boiler pressure is also required, and must be connected to the boiler. The pump is connected to the boiler and a water supply, and all other outlets except one must be sealed. The boiler is then *completely filled* with water so there is no air space left at all, and the remaining space, through which air has been escaping while the filling took place, is sealed off. Water is pumped into the boiler until pressure reaches twice the working one, and this pressure is retained for about a quarter of an hour – if it is necessary to give a few strokes of the pump in order to maintain it, that is fine. The boiler can now be deemed safe.

A leaky boiler does not mean an unsafe one, and in full-size practice nearly all, if not all, boilers had a leak somewhere or another; but it is desirable that there are no leaks in small boilers such as these. They do not have the capacity to supply steam to an engine as well as to send it to the atmosphere via a leak. As far as leaks are concerned, if they are only very minor ones, leave well alone; they will seal themselves when the boiler is heated and will not cause any problems. If they are bad ones some more soldering must take place in order to seal them, otherwise the required pressure when working

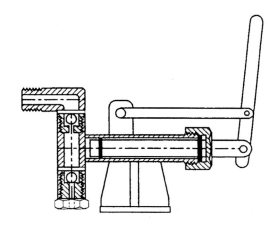

Typical hand pump layout

The pump can be fabricated from brass tubing, the proportions as per the drawing. A maximum 8mm (5/16") dia. ram should be used. The end gland is sealed with an "0" ring and the ram has a maximum lift of 1mm (1/32").

Typical hand pump layout.

will never be reached or if it is, it will not be possible to maintain it.

Those readers not having the facilities for testing, or who perhaps do not want to go to the trouble of making a hand pump and fitting a gauge in order to test a boiler, can go ahead and use it, providing only a specified burner is used. It is advisable, though, to do so without lagging it, so that if there are any leaks that need attention they can be seen. It can be lagged once you are sure it is going to do the job.

13 Boiler Fittings

A number of the components will be required for use when connecting a boiler to an engine, so the following items can be classed as universal, as they will work with all the boilers and all the engines – providing, of course, they are made with matching threads.

UNION NUTS

Suitable union nuts can be bought from model-engineering suppliers; they are very easily made by anyone with a lathe, and the following description will help. The nut is made from hexagon material, and brass will do nicely for this job. Put a length in the lathe chuck, face off and then use a centre drill to start with; drill the tapping size to the required depth, followed by a clearance drill for the pipe that is being used, which is passed to a depth equal to the overall length of the nut. Using a 'D' bit, flatten the end of the tapping diameter, and then tap it. Most commercially made taps bought for this job are unsuitable, as there is too much of a taper lead to allow the thread to go to the end. It will be a case of grinding a tap until there is a flat end at full thread diameter; it will

then tap right to the end of the hole. All that need be done then is to part off, and the nut is made.

NIPPLES

Although nipples can also be purchased, they are of a different type to the ones about to be described, because they are coned: in other words, they have a taper throughout most of their length. As a result of this they are larger than the type suggested, and the fitting into

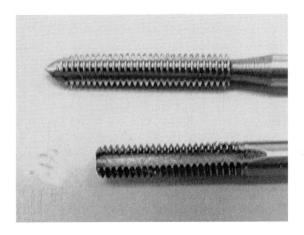

The photograph shows two taps, the lower one having had the end ground away so that it will make a thread right to the end of a union nut. The other is a standard plug tap that has been ground, and it can be seen quite clearly that to use it would mean terminating the thread some distance from the end of the nut.

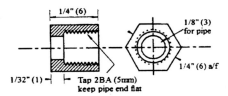

Exploded view of union nut.

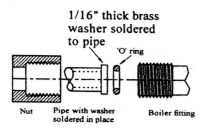

1/16" thick brass
washer soldered
to pipe

'O' ring

Nut   Pipe with washer
soldered in place   Boiler fitting

Exploded view of arrangement for connecting piping.

which they go requires a matching recess for the cone; this means the use of a slightly larger-diameter thread that does not look quite as neat as those chosen for these projects.

To make the nipples, simply take a length of brass rod of a suitable diameter, put it in the three-jaw chuck and, using the tailstock drill chuck, drill a short length of the rod clearance size for the pipe. Make the hole an inch or so deep and then part off ⅟₁₆" (1.5mm) slices of the

metal. Make sure there are no burrs either on the periphery or in the central hole: the nipples are soldered to the length of pipe, but make sure they are fitted squarely.

REGULATORS

It has already been explained that some items will be common to most or all of the models, and they are dealt with separately in this chapter; included in this category are regulators and reversers.

The regulator is nothing more than a tap that turns the steam on, rather like the ordinary domestic tap, except that it turns on water instead of steam. The domestic water tap works by screwing a plug into a hole to prevent or allow the passage of water, and suitable screw-down taps designed for use on miniature boilers can be obtained from any supplier of model engineering tools and equipment. Regulators do, of course, vary to some degree in their complexity, but no matter how complicated the

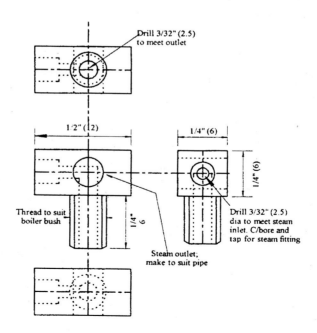

Drill 3/32" (2.5) to meet outlet

1·2" (12)

1/4" (6)

1/4" (6)

Thread to suit boiler bush

1/4" 6

Drill 3/32" (2.5) dia to meet steam inlet. C/bore and tap for steam fitting

Steam outlet; make to suit pipe

Regulator body (twice full size).

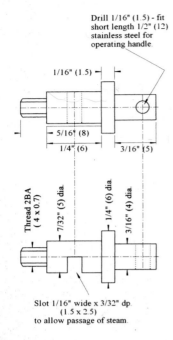

Drill 1/16" (1.5) - fit
short length 1/2" (12)
stainless steel for
operating handle.

1/16" (1.5)

5/16" (8)

1/4" (6)

3/16" (5)

Thread 2BA
(4 x 0.7)

7/32" (5) dia.

1/4" (6) dia.

3/16" (4) dia.

Slot 1/16" wide x 3/32" dp.
(1.5 x 2.5)
to allow passage of steam.

Regulator plug (twice full size).

Nut: thread to suit body, ⅜" (10) a/f x ¹⁄₁₆" (1.5) thick.

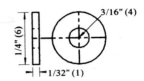

3/16" (4)

1/4" (6)

1/32" (1)

Soft washer or gasket: make from neoprene or similar material. Also required a similar item with hole ⁵⁄₃₂" (5) dia.

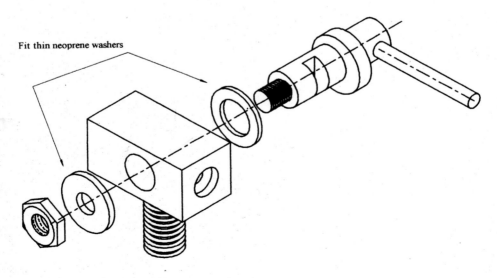

Fit thin neoprene washers

Exploded view of regulator. Not to scale.

particular design might be, in the end it is only going to turn the steam on and off. The main difference will be that a complicated type will possibly allow a graduated amount past, whereas with a simple one it will be a case of all on, or all off. Not that such a thing would matter too much to us, because in most cases full power will be the order of the day; but it is nice to have a degree of control that a well made regulator can supply.

There is no need to dash out and buy a regulator, as they can very easily be made in the home workshop, not being in any way complicated: in the time it takes to go out to order or get one from a supplier, a perfectly good one can be made.

The body can be made from square or hexagon brass or bronze, the latter material being preferable; it consists of nothing more that a short length of material with three holes drilled in it: the one that goes across it is to accept the plug that turns the steam on and off; a second is connected to the steam outlet of the boiler, the third is the outlet from the regulator. The drawing shows the bottom hole fitted with a threaded pipe ready to connect to the boiler; this arrangement may not necessarily suit every-one, and some readers will possibly wish to place the device away from the boiler and will therefore want a pipe connection there instead.

The shape of the plug can be clearly seen in the drawings, and it is essential that it is a good fit in the hole in the body, as otherwise steam will leak. It is the groove that regulates the movement of the steam; with the plug in the closed position nothing can pass, then as it is turned, the groove slowly provides a path for the steam to pass through. The handle should be made of stainless steel to prevent too much heat being transferred along it and burning one's fingers when operating it.

REVERSING VALVE

To reverse an oscillating engine we simply change the port into which the steam is admitted, and to do this it is necessary to

A disk reverser of the type that has been described. It has been mounted on a stanchion for mounting on a base, but could just as easily have been fitted to an engine directly.

transfer it from one pipe to the other. This can be done in a number of ways, but in keeping with the title of this book the simplest method is shown. Because it is the easiest method it is also not the most efficient, but it does its job, which is all we can ask for; the only problem is that if the two mating surfaces of the body are not made carefully, the device will leak a certain amount of steam.

This simple reversing valve consists of just two discs of bronze or brass, one of which has four holes, which are connected in sequence to steam, port one, and exhaust, port two. The other disc contains two slots, and in one position the slots will connect the steam to port one and the exhaust to port two. By rotating this disc the steam becomes connected to port two and the exhaust to one, thus reversing the

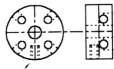

Tap for fitting

Fixed disk for reversing valve 3/4" (20mm) dia.
Four holes 3/32" (2mm) dia,on 9/16"
(14mm) dia. pcd Stanchion or other
holding device 7BA (2mm)thread .
Pivot hole 1/8" (3mm) dia. Holes for
steam passages may emerge on
periphery or back of disk.

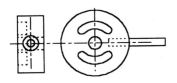

Moving disk, 3/4" (20) dia.
Slots to match holes on fixed disk. Handle
threaded 5BA (3mm) 7/32" (2.5) dp.

Fixed and moving disks for a reversing valve.

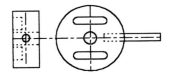

*Alternative method of cutting slots
in moving plate.*

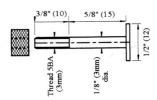

3/8" (10)   5/8" (15)

1/2" (12)

Thread 5BA
(3mm)

1/8" (3mm)
dia.

ABOVE: *Pivot pin – fit with spring.*

RIGHT: *General arrangement of disk type
reversing valve, showing connections. Although
shown mounted on a stanchion it can also be
fitted directly to a boiler. Measurements can be
altered to suit individual requirements.*

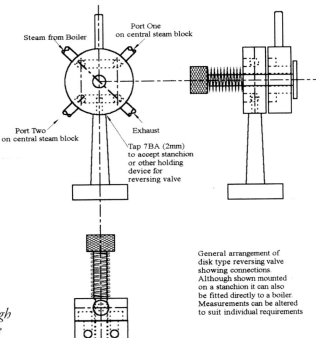

Steam from Boiler

Port One
on central steam block

Port Two
on central steam block

Exhaust

Tap 7BA (2mm)
to accept stanchion
or other holding
device for
reversing valve

General arrangement of
disk type reversing valve
showing connections.
Although shown mounted
on a stanchion it can also
be fitted directly to a boiler.
Measurements can be
altered to suit individual
requirements

direction of rotation of the engine. The slots
ideally should be made in a radius, as doing so
gives a better clearance; however, it is not easy
to do this without access to such equipment as
a milling machine and a rotary table. The only
way for the average person with limited
equipment is to scribe a line along the radius
and then drill a row of holes along that. The

holes have to be close enough to break into
each other, and then the radiused slots can be
finished with a file. Sounds easy, but don't
believe it, because it is a very difficult operation
indeed.

An alternative method is to take the slots
straight across the face of the disc. Doing so
makes little difference to the operation of the

device, but again we run up against a bit of a snag. The slots will now be dangerously close to the edge of the pivot hole, and a certain amount of space is required to make the reverser steam tight. It is therefore necessary to ensure that the diameter of the disc and that of the pivot hole are of such a size that there will be clearance.

The device is finished with a pivot, and the two parts are held in place with a spring, which is secured and adjusted with a nut in a similar manner to the way the cylinder is secured on the engines. The only other requirement is a small handle that is fitted in the disc containing the slots. The reverser will also act as regulator, because when the handle is set in suitable positions, no steam at all will be admitted, so there is no necessity to fit both regulator *and* reverser: either will do.

SAFETY VALVES

Safety valves have been standardized, and the springs from ballpoint pens will usually prove ideal for use in them. Rustless springs are desirable, and can be wound by the builder by simply pulling a length of phosphor bronze or stainless-steel wire round a former, held in the lathe chuck. The wire is held taut in a pair of pliers, and the lathe rotated by hand; the wire must be kept in tension all the while. The wire

can be trapped between the chuck jaws and the former to keep it in place while the spring is wound. This is not the correct way to do it: you should drill a hole in the former and bind the wire through that, but with such a small-diameter former it is not very practical, and in addition would almost certainly weaken the former to the point where it would break. The lathe is rotated by hand, and the wire kept in tension by holding it taut with a pair of pliers. The coil will open slightly after winding, so allowances must be made for that to happen. Springs made in this way are frequently of uneven spacing, but this does not prevent them working.

Those wishing to make something a little more sophisticated can cross-drill a close clearance hole in a short length of brass or steel rod, then drill it centrally in the lathe and solder the spring former in that. Another close clearance hole is drilled in a block of hardwood, which is mounted in the tool post and the tension again taken with a pair of pliers. Instead of rotating the lathe by hand, it can be set to self-act, with the change wheels set to give a feed of about twenty to the inch (or 2mm). The spring will then wind under power, and the coils will be evenly spaced.

The wire can be cut off with wire cutters, and the end that stands proud levelled with a small carborundum disk in a drilling machine.

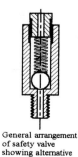

General arrangement of safety valve showing alternative steam release.

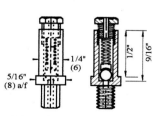

Safety Valve. General arrangement and body details. Bottom thread to suit boiler, central hole 1/16" (1.5mm dia. internal bore 5/32" (4)

Adjusting Nut, Thread to fit body 1/4" (6) long, drill 3/64" (1mm) dia. hole centrally through length of nut, for steam escape.

Imperial and metric drawings of guide rod. Use 1/8" or 3mm ball. Radius of guide bottom to fit.

Safety valve: make from brass or bronze. Use spring made from 28 gauge (0.35mm) bronze or stainless steel wire, nine turns over length.

Safety valves should always be tested before use; this can be done with a car foot pump that has a pressure gauge attached, by simply passing the air into the valve and noting the pressure at which it releases. It can also be done using the air pressure hose at a garage. Any adjustment should be made to the release before the valve is fitted to the boiler.

PRESSURE GAUGES

No doubt most readers will have noticed the complete absence of pressure gauges, and this is quite deliberate because using the spirit burners specified there is no danger at all of over-pressurization. If one is thought to be necessary or useful, they can be purchased, but when doing so a siphon should be obtained at the same time. The siphon prevents a sudden surge of pressure from damaging the Bowden Tube, which is the heart of the gauge and is a worthwhile investment. It is quite possible to make pressure gauges in the home workshop, but doing so is not within the scope of this book.

WATER-LEVEL PLUGS

Rather than fit water gauges, the use of plugs is suggested. No doubt some people will prefer to buy or make a gauge, but gauges can create problems of their own with air locks causing false readings, particularly where small boilers are concerned. The test plugs take two forms: a standard screw-in plug that is completely removed to allow the level to be assessed; or the slightly more sophisticated version with water-ways drilled in it, that need undoing by about a single turn to allow water to drip out rather than come out at full pressure as is the case with the simpler version. Drawings for both types are shown, and it is up to the reader to choose which type they wish to use.

TURRETS

A turret is a device for splitting the steam supply to several different controlled outlets, and the easiest way to think of one is to visualize several regulator valves stuck together. In fact what we do is to make a single body of extra length, with provision for two or more valves; so there is nothing very complicated about it. A single drawing of a completed four-outlet device is shown; any measurements that might be required can be taken from the drawings of the single regulator valve.

IN GENERAL

When making fittings of any type, bronze is the material to be preferred, as it does not suffer from de-zincification, as does brass. There are a number of free cutting bronzes available that are suitable for this sort of work, and it is wise to discuss a suitable material with a metal dealer or model-engineering supplier. Some bronzes are very difficult to work with because they

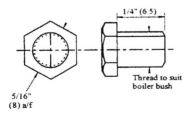

Plain plug for checking water level when filling a boiler.

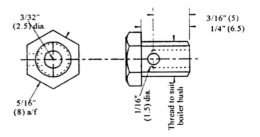

End plug with relief passage as alternative to plain plug.

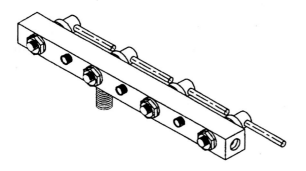

Four regulators have been connected via a single block to form a turret with four outlets. The position of the outlets can be altered to suit the particular requirements. The turret can be used to connect one boiler to four engines. The number of outlets can be increased or decreased, depending on the number of models.

quickly generate heat, and this causes the drill to stick in them, with almost inevitable breakages; they also quickly cause tools to go blunt. Bronze is an alloy of copper, and one of the metals that it is sometimes mixed with is aluminium: however, this type of bronze is not suitable for making fittings, as it cannot be soldered successfully.

If in any doubt about the quality of available bronze, make the fittings from brass; they will last a number of years, and can easily be replaced. However, if brass is used, check it at

six-monthly intervals to see if there is any deterioration of the metal, particularly in or around the threads that hold it to the boiler.

THREADS FOR CONNECTIONS

The connections have as far as possible been standardized, and the threads chosen are in imperial 2BA and metric 4 × 0.75mm. To model engineers used to working with the standard ME threads, 2BA may seem a strange choice – but the pitch is within a gnat's whisker of the ME standard of $\frac{3}{16}$" × 32, and anyone with ME threading equipment can use $\frac{3}{16}$" × 32, or preferably $\frac{3}{16}$" × 40. Imperial 2BA has been chosen as the thread for which most people are likely to have taps and dies. Those readers using the American system can use UNF 10; again, the specification is very close to that of $\frac{3}{16}$" × 32.

The metric thread of 4 × 0.75mm is in the Metric Fine series, which in many ways is not quite as good as the ME series; but it does work, and once again it is a thread for which taps and dies are possibly already available. The finer thread in the series of 4 × 0.5mm would be preferable, and if anyone finds it necessary to purchase taps and dies, then 4 × 0.5 is recommended. Some readers may therefore feel it worthwhile to purchase the equipment to use the finer thread, even though they already own the taps and dies for the coarser one.

14 Boiler Number One: Pot Boiler

This is absolutely the simplest type of boiler, and is identical in most ways to those that are supplied with steam toys – which incidentally have hardly changed since the first ones that were sold. It is known as a pot boiler, because that is literally all it is: a pot, used in the horizontal position. It consists of a length of seamless tube with a plate at each end and three holes that are fitted with bushes. The end plates are flanged over a short piece of round bar machined to the correct diameter, and they are then soldered in position. But before doing so, two holes are drilled in the top and another in one of the end plates, for the above-mentioned bushes. (See Chapter 16 for details.)

The drawings show the boiler as made from 2" (50mm) copper tube, but any size will do,

with one slight proviso: if a larger diameter tube is to be used, then a copper or bronze stay should be passed lengthways through the end plates, and secured with nuts or silver soldered in place. The reason for this is that we rely on the fact that tubular material is inherently stronger than sheet, and of course the ends of the boiler are sheet material and liable to bow outwards under pressure. The specified gauge of copper at the diameter shown, and using a pressure of no more than forty pounds per square inch, will not react in this way, but larger sizes will, and the central stay prevents it. Note that the ends are of a heavier gauge copper as a precaution against the possibility of them blowing out under pressure. It is thought highly unlikely that any reader will use tubing

This is about the easiest possible type of boiler to build: known as a 'pot boiler', it consists of nothing more than a copper tube and two flat plates at the end. Two bushes are used, one to check the water level, the other to fill the boiler and hold the safety valve.

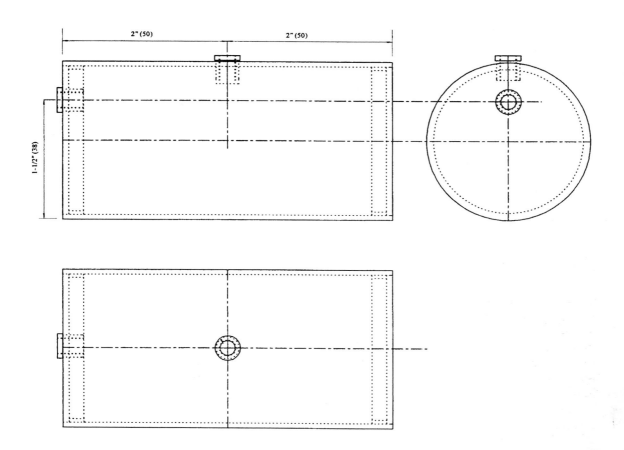

Number one boiler. 2" (50mm) dia. Make from 20 gauge (1mm) seamless copper tube. Flanged end plates 16 gauge (1.5mm) thick. Use standard bushes. Alternative may be used on top.

much above 3" (75mm), but should anyone decide to do so, then a number of longitudinal stays will be needed.

BOILER BUSHES

The three bushes should preferably be made from bronze, but if that is not available then brass will do. They can easily be made on a small lathe, and if the specified screwing equipment is not available, any similar size will do. The pieces that fit in the bushes will, of course, have to be of the same thread. The bush

in the end plate is there so that the water level when filling up can be gauged. The plug is taken out, and as soon as the water drips through the hole there is sufficient in the boiler. *Under no circumstances should the plug be removed while the boiler is in steam or being heated.*

One top bush takes the steam to the engine, and nothing more than a connection to a pipe is necessary. However, no doubt most readers will prefer to fit a regulator valve, as otherwise one has no control over operations, and then take the steam from that to the engine. The other bush in the top is for a safety valve that should be set to release pressure at 30lb per square inch maximum. In fact all the engines will work on 20lb per square inch or less, and it would be advisable to set the safety valve to release at a little above that.

What form the stand for the boiler takes must be left largely to the constructor. Some people will be quite happy with a simple stand and the boiler fixed on top of it; others may wish to make a totally enclosed arrangement that gives the appearance of an engine house. Whether a stand or something more complicated is chosen, it is essential that a good quantity of air can penetrate to the burner, otherwise it will not work properly. Whatever the method, air must be able to circulate freely in order to get maximum efficiency from the burner.

LAGGING

The performance of the boiler can by considerably improved if it is lagged, and there are a variety of ways of doing this. Possibly the most efficient is to enclose it entirely (except for where the bushes are) with a heat-resistant material of some sort; at one time asbestos would have been the preferred medium, but that is now forbidden. A good alternative is fibreglass matting as sold by motor accessory shops; it is available in a variety of thicknesses, and can be cut to size and shape easily with a pair of scissors. It can be bound to the boiler with wire, and then covered with metal sheeting. Soft drink cans are ideal for this, and are readily available. If a steel one is used, it can be soldered to make a join at the bottom of the boiler; the aluminium ones cannot readily be soldered (unless one is proficient in this). The best way to fit them is to cut thin strips from a steel can, pull them tightly round and solder underneath. The ends can be dealt with by flanging them using the same method as was used to make the boiler plates, and tucking the flanged ends under the rolled main lagging.

An alternative method of lagging is to use thin wooden strips, an idea that finds considerable popularity amongst model makers. Various types of thin wooden strip can be bought at shops specializing in model boat or aircraft construction. The wood is simply laid over the boiler and secured by bands. The bands cannot be soldered as the wood will get charred, so their ends are bent at right angles and they are drilled to accept a small nut and bolt. Lagging the ends with wooden strips is neither practical nor pretty, and in fact it looks entirely wrong. The ends are therefore either left unlagged, or metal plates can be flanged and the flanges tucked under the wooden lagging.

15 Boiler Number Two: Basic Vertical Boiler

This boiler is just the tiniest bit more difficult than Number One, but the extra work involved in making it pays for itself by the extra efficiency that is gained. The drawing shows a tube of the same diameter and thickness, and this is no accident: it is a handy size and easy to work with, and when it was known that this book was to be published, a length was purchased and cut to suitable lengths for the boilers. Instead of lying horizontally as Boiler One, this is a vertical boiler – not the sort of thing one would find driving mill engines, but popular in factories and similar establishments where steam was the driving force and room was at a premium.

THE BARREL

The barrel is started in the usual way, by cutting to length and squaring the ends. It has two holes in it, spaced as shown, and both are bushed. The plugs that fit in these bushes are made a little differently inasmuch as they are drilled centrally along the thread and then cross-drilled, giving an 'L'-shaped passage; details of this type of plug are given in Chapter 13. When the plugs are partially unscrewed, steam or water can escape via the passage, which is, of course, the reason for fitting them. When the boiler is working, if the top one is

opened sufficiently for the cross-drilled hole to clear the bush, there should be an escape of steam; however, if the lower one is opened and there is an escape of steam, it is time to close things down as there is insufficient water in the boiler. If, instead of steam, water leaks out, then everything is fine and the boiler is working

The completed boiler, showing the bushes and the central tube that is used to carry away excess heat from the burner and to assist in heating the water.

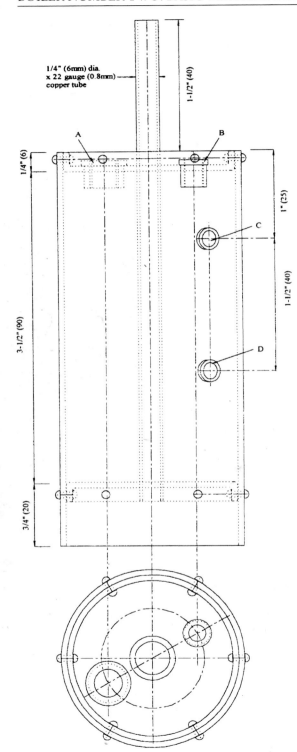

1/4" (6mm) dia.
x 22 gauge (0.8mm)
copper tube

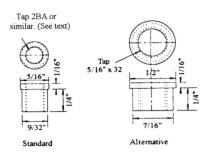

Tap 2BA or
similar. (See text)

Tap
5/16" x 32

Standard

Alternative

Bushes in imperial form. Fit as per metric. All bushes from brass or bronze.

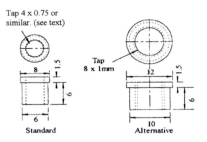

Tap 4 x 0.75 or
similar. (see text)

Tap
8 x 1mm

Standard

Alternative

Standard and alternative bush metric forms. Fit standard size at B-C-D and alternative at A if preferred to alternative.

correctly. The top one can also be used to check the water level when filling up.

This type of plug has obvious advantages over the plain one. The latter cannot be used when the boiler is actually operating, because the amount of steam released would be dangerous and would easily burn anyone nearby. The plug with the hole ensures there is a very limited steam release. However, care must still be taken not to unscrew the plug too far so that it comes right away. (See Chapter 16 for bush details.)

Two flanged plates are made, and each has a

Number two boiler. 2" (50mm) dia. Make from 20 gauge (1mm) seamless copper tube. Flanged end plates 16 gauge (1.5mm) thick.

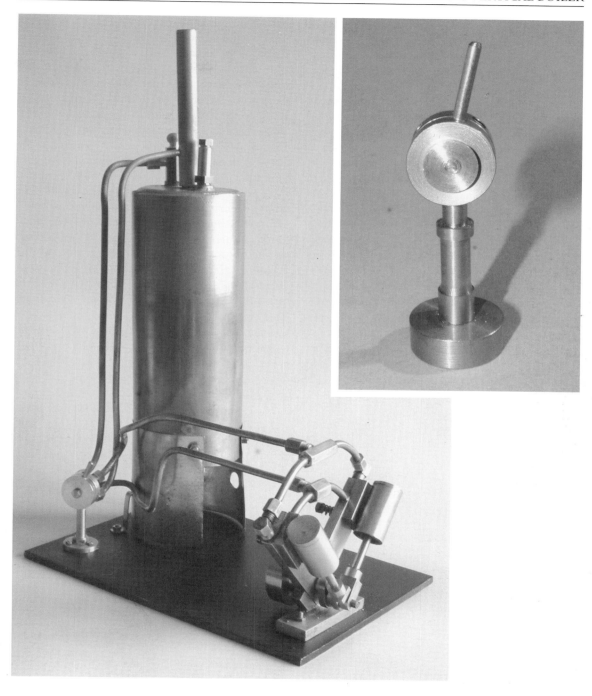

The vertical boiler seen here connected to the vee type engine via a disk regulator.
A stand has been made for the boiler by simply wrapping an old tin round it.
Before doing so ventilation holes were punched in the tin.

central hole drilled in it; the one at the top has two extra holes, both of which will be fitted with bushes, one for the safety valve, the other for the regulator. The standard size used for the safety valve will do for filling as well, but a larger-sized bush is also shown as an alternative: using this will allow the boiler to be filled more quickly, but it will also mean that an adapter will be required in the form of a ring, with a thread for the bush and a tapped hole for the safety valve. The central hole in each plate is for a length of tube to go through, which acts as a chimney; it releases the surplus heat from the burner and also imparts extra heat to the water, thus giving the boiler better steaming properties.

Soldering the boiler is done in exactly the same manner as with Number One, but note that the bottom flange plate is a little way inside the shell or barrel. This leaves a short length of the barrel available for screwing a suitable stand to.

Lagging the boiler can take the same form as with Number One, except that it is, of course, not possible to lag the ends. The bottom end is there to be heated and so needs no further attention; however, it is best to cover the top one by cutting a circular plate with a hole to clear the central tube or chimney, and smaller ones for the fittings. Dropping that on will stop some heat from escaping, and will provide a nice finish to the boiler; the plate can be flanged if made from a suitable material, and this will give an even better appearance.

A piece of thicker tube slipped over the one that has been soldered in place can be used as a chimney; the length will depend on individual taste. A nicely machined cap will finish the whole thing off. The chimney will hold the cover in place, and that will also be secured when the fittings are screwed in, as they will press down on it. A round burner is required, and the little one that was used for Tin Can Tommy, made from a lozenge tin, was found to work reasonably well.

16 Horizontal Boiler with Water Tubes

This little boiler is considerably more efficient than the simple pot boiler, and it must be said that a little more time and care has to be taken over building it. That time will be well spent, however, as the boiler is not only compact in size but is able to raise steam very quickly. It is made from the same 2"- (50mm-) diameter copper tubing, with the end plates from 16-gauge or 1.5mm copper sheet.

We start in exactly the same fashion as for the pot boiler, by trimming the shell to length and making a pair of end plates that are flanged over. The former for the flange plates need be nothing more than a short length of round bar of suitable diameter. When making them, do remember that it is far better to heat the copper several times and bend it over gradually, rather than try and belt it to shape with a few lusty hammer blows.

Once the plates are shaped and finished to size, which will probably mean filing the edges, the hole that is shown on the drawings, which is for a bush, can be drilled. The usual precautions when drilling copper should be taken: an easy action, with a drill that has been slightly blunted, regular clearing of the swarf, and the drill running at a slow speed. Do not apply any form of cutting oil, as it can become impregnated in the copper when that heats up, and it will prevent the bush from being able to be soldered in place.

Although shown with the flanges inside the shell and the edges pointing inwards, if the reader so wishes the plates can be made to fit outside the shell; or they can be fitted inside, but with the flange pointing out. Everyone will have their own idea on the best way of fitting them, but in fact there is no correct way, although traditionally the flanges are placed inside the shell, pointing inwards.

Three tubes fitted along the bottom of the boiler give an increase in heating, as the limited amount of water that they contain is heated prior to the main contents. The result is a boiler that creates steam faster then the ordinary pot boiler.

It will be seen that there are three tubes protruding from the base of the shell; these carry water, which heats up more quickly than the larger amount inside the boiler. The size of pipe chosen for this job was ⁵⁄₃₂" (4mm) diameter, and was selected as being the largest practical size when it comes to shaping, unless special equipment is available. In fact ³⁄₁₆"- (5mm-) or even ¼"- (6mm-) diameter tubing would be better, but getting it to the required shape is far from easy.

The best way to bend the tubes to shape is to proceed as follows: thoroughly anneal them, solder a piece of brass or copper on one end, and fill the tube with water. Put it in the freezer until the water becomes ice, and then make the bend. The ice will stop the tube from kinking, and should the tube-work harden during the process it can easily be re-annealed and the process started again until all is well.

Mark out, drill, and if possible ream the six holes required for the tubes; the holes need to be quite accurate, otherwise the solder will run through and will not seal the joints. The next task is to bend the tubes to shape, and when that has been completed, to get them to fit in the holes. The tubes should not be fitted too close to the surface of the boiler, because we want the heat to spread right round them. Because they need to be comparatively close, they will be entering the holes in the boiler at a slight angle, created by the tube curvature. To sort this out, take a piece of mild steel the exact diameter of the holes in the shell. Push it in a hole, and then gently lever it towards the centre of the shell. The copper will kink very slightly, which is what is wanted, and a bit of careful pulling will get the holes to the required angle to accept the tuning. It will almost certainly be necessary to trim the tubes to size; the maximum amount inside the boiler should be ⅛", or 3mm. As the ends have not yet been put in place, it should be possible to do some trimming off after the tubes have been soldered in. Keep the amount on the inside to the minimum, as the longer the internal part, the more water is required to cover them.

Two holes for bushes are required at the top of the boiler, and these can be drilled at this point. They should be a good fit for the bushes, although the fit is not quite as critical as it was for the water tubes, since the lip on the bush will always act as a barrier to hold the solder

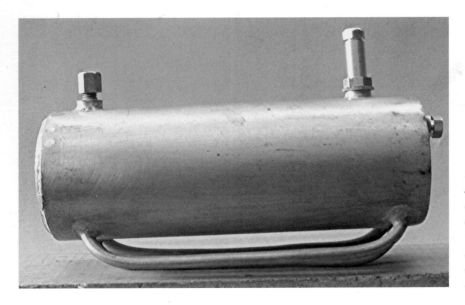

The finished boiler; note the water tubes are comparatively close to the bottom of the drum, but with sufficient space to allow heat to circulate right round them.

The boiler, mounted on a stand and connected to "Millie" makes a fine model. To enhance the appearance the assembly has been mounted on a wooden plinth with a simulated stone effect on the edges. The simulated stone is actually only sand paper with the stone edging marked in pencil.

and prevent it running inside. Obviously they need to be in line, and a useful idea for scribing a line accurately along the length of a piece of tube is to use a length of angle as a guide. It seats itself nicely along the curvature, and remains in place. Trying to use a ruler is most frustrating, as it will fall about all over the place.

When these two holes have been made, the end plates can be soldered in position, followed by the bushes, and then the boiler is complete. A word about the bushes will not come amiss here: they should preferably be made of bronze, with brass being used only if it is absolutely necessary to do so. When making them, it is a good idea to run a tap about three-quarters of the way through, and then finish the threads after the bushes have been soldered in place.

Frequently it will be found that the bush is not completely square, and nobody wants a safety valve leaning over at an angle. The answer is to thread a length of mild steel rod and screw it in the partly finished hole, then

check it with an engineers' square. If it is at an angle, pull it into the correct position. The copper should be nice and soft, so it will move quite easily, and because the thread in the bush has not been completely finished, it will come to no harm. Once satisfied that the bush is square, run the tap right through to complete the thread.

The bushes have been specified as the same for all boilers, so any fittings will be interchangeable, but this has meant using quite small diameter threads and while they are fine for the fittings, it is a little difficult to fill the boilers through the holes. It would be better, therefore – and assuming that suitable threading equipment is available – to fit a larger bush and make an adapter for the fitting. The adapter is simply a piece of bronze that screws into the bush and accepts the fitting. It is best made from hexagon material so that it can be both tightened and loosened with a spanner, which will ensure it is steam-tight when in place.

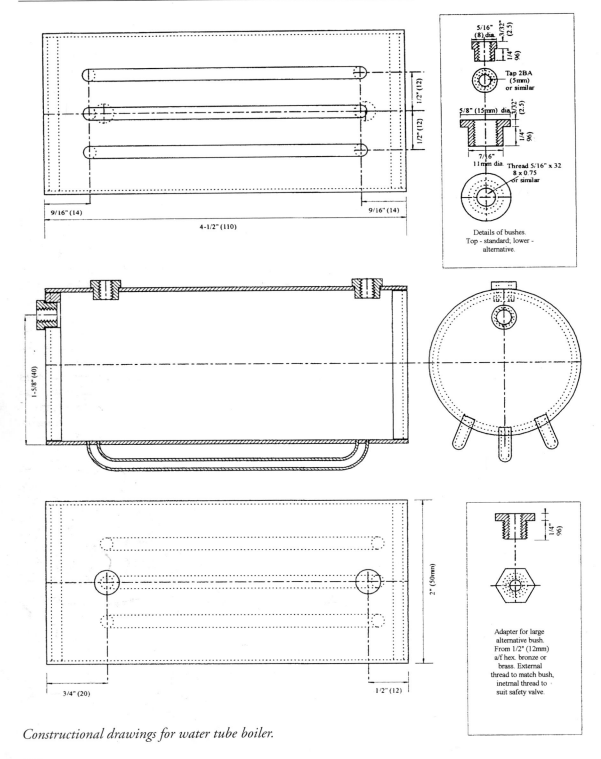

Constructional drawings for water tube boiler.

17 Vertical Boiler with Flues

This is a vertical boiler made on the same principles as the previous one, the only difference being that there are a number of vertical flue tubes, rather than the single one. The central tube remains, and is left standing proud to give the appearance of a chimney, which can be fitted with a brass cap if one wishes. The extra tubes, usually referred to as flues, draw additional heat through the water giving greater efficiency, although this is at the cost of the reduction of water capacity.

The easiest way to drill the holes for the extra tubes is to make suitably spaced holes in the former used for making the flange plates. Once those are completely shaped, at least as far as hammering is concerned, a drill can be passed through the holes in the former; this ensures that even if there has been a slight error when marking out the tube positions, at least both flange plates will be identical.

As usual, rivets should be used to secure the end plates during soldering operations, and the flue tubes should protrude no more than ⅛" (3mm) above and below the plates; the height of the one left as a chimney is a matter of

The former used to make the flange plates was also used for the hole spacings. It is easier to mark out and make the former than it is to mark out and drill the holes in the end plates. The thickness of the former allows it to be held in a machine vice for the purpose of drilling, whereas the end plate must be clamped down or hand-held. The former is only a piece of mild steel bar of the correct diameter with a suitably radiused edge.

personal choice. If the tubes are made slightly different in length, while it might spoil the appearance of the boiler just a tiny bit, they are easier to locate through the holes as they can be pulled through singly rather than trying to get them all in place at the same time. If the uneven length is at the top end, they will be covered anyway and so ultimately will not notice. Bushes are fitted on the barrel for the usual plugs used to gauge the water height, and there are also two on the top of the boiler, one for the steam take-off, the other for a safety valve.

The extra flue tubes mean that there is extra heat coming out at the top of the boiler, and while this is fine as far as making steam is concerned, it makes life rather uncomfortable

for the operator. It is therefore necessary to fit a cover, which is best made of brass or aluminium; steel will rust rapidly owing to the condensation created by the burner. There are two ways of doing this: firstly, a plate can be flanged in a similar manner to that used to make the end plates, and it can be made so that it just slips inside the boiler shell. A slightly easier way is to make a disk and solder a flange to it using a strip of brass, again making it so it slips just inside the shell. Alternatively you can take the easy way out and have just a flat cover in the form of a metal disk.

The former can again be used to ensure the holes in the cover are in the right place, but only spot through it, do not try and drill the cover, because in those circumstances it is certain to snatch, and not only will that ruin the cover, it is also likely to result in injury to the operator. Subsequently the cover can be clamped down while drilling operations are completed.

The cover will need three holes, one for the chimney to pass through, and the others for the safety valve and steam valve. Again, it is a good idea to make one bush slightly larger as an aid to filling the boiler. It may not suit some readers to use the larger bushes, as together with the adapter they do look rather unsightly, in which case a small syringe will be needed for boiler-filling operations.

The photograph shows the top of the boiler, and the tubes and bushes can be seen quite clearly. The bushes have two adapters fitted: on the left the adapter is tapped to accept the safety valve, on the right it has an outside thread to accept the regulator or on/off valve. The extensions are necessary because the boiler bushes are below the top of the shell, and without them it would not be possible to fit a cover and also the safety valve and regulator.

A plain disk used as a top cover; it not only tidies things up, it also prevents heat loss to the atmosphere. The holes for the bushes are slightly oversize to ensure there is plenty of space to screw in the fittings.

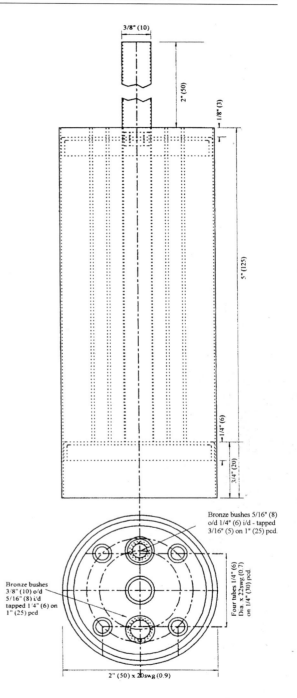

ABOVE: The boiler top, showing the safety valve and regulator in place. Note how the safety valve fits into a hexagon extension bush, the head of which, together with the central tube, keeps the plate in position.

RIGHT: Boiler number four. Similar to number two but with additional fire tubes.

The regulator, or on/off valve that has already been described, looks very unsightly on this type of boiler, but there is no reason why it should not be adapted for use. All that is needed is to move the inlet and outlet connections so they are in line with the body, and the job is done.

18 Boiler Number Five: Double-Barrelled Boiler

This little boiler is ideal for those who do not have the facilities for silver soldering larger ones, but prefer not to use soft solder; it can be easily silver soldered with a small DIY-type blowlamp. The design is a simplified version of some full-size boilers; the model shown in the photographs was made from 28mm copper tube, a size once common in household plumbing but now largely replaced with plastic. It is still fairly common at scrap merchants, and many plumbers who work on modernizing old houses will have some. The piece of tubing that was used came from an old house and therefore was not in pristine condition: it had paint on the outside and a coating of lime residue or fur on the inside. The paint was removed with a blowlamp, and the inside cleaned by leaving the copper to soak for several days in a solution of citric acid, made up as about two tablespoons of acid crystals to a bucket of water.

As not all readers will wish to go hunting around scrapyards for the necessary material, a cheap alternative would be to use 22mm tubing, which is still in common use. It can be bought at plumbers' merchants or DIY stores in full lengths, or off-cuts can usually be obtained from any plumber – frequently they are only too glad to give it away! Another alternative is to use the standard 2" (50mm) that is specified for all the other boilers, but this will mean that the domestic type of blowlamp is unlikely to cope with silver soldering, and that the finished boiler will be quite large. Naturally the use of 22mm will reduce the capacity of the boiler, thus giving the engine a shorter running time, and if this poses a problem then there is no reason why an extra inch or 25mm should not be added to the length. Most model-engineering suppliers stock 1"-diameter tubing and sell it in short lengths, and this could be another source of supply; unfortunately the thickness is often greater than we need, nevertheless it would almost certainly do the job. With small diameter tubing such as we are dealing with for this boiler, working at relatively low pressures 22 swg (0.7mm) wall thickness is quite strong enough. If the copper is too thick, the heat supplied by the burner will be used to heat the copper, instead of the water inside.

The boiler is only really suitable for silver soldering together, and before doing so, each part must be thoroughly cleaned: the easy way to do this is to put it in a solution of citric acid and leave it in the solution for four to five hours at least, and if possible longer. After each soldering operation it should be allowed to cool a little, and then plunged into cold water to finish the cooling process; afterwards it can be put into the acid solution for cleaning.

Silver soldering is carried out in exactly the same way as soft soldering, but the temperatures required are considerably higher; this means that one of the larger DIY blowlamps is essential for heating. The work after cleaning must be fluxed, and as all suitable flux invariably seem to be sold in powder form, it is advisable to mix a little flux with some water until a fine creamy mixture is obtained, which can be spread on the parts around where the joint is to be made. The flux will bubble before the work is hot enough to apply solder, and in general the job will have a dull red glow when it is at a suitable temperature. As with any type of soldering, the solder must melt on the work and not as a result of the flame being played on it, and so it should be applied at the side of the point of heat and allowed to flow, rather than attempting to apply it directly under the heat.

Once silver solder has been applied to the work and allowed to cool, it will not melt again at the original temperature, and this effect can be used to our advantage, as it is quite possible to apply silver solder near a previously made joint without causing any damage to the original one. Although there is a small degree of extra temperature required to re-melt *soft* solder, it is an insignificant amount, and trying to apply more solder near a joint that has already been made will almost certainly result in the original one coming undone.

Make a start by ensuring the ends of the two large tubes are square; anyone with a large lathe can do this with a turning tool, though if it is to be done in the lathe, no more than ¼" (6mm) should be protruding from the chuck: copper is an awkward material to machine and easily snags in the tool, so if there is an overhang of any length, it will mangle the copper. If the lathe will not accept the diameter without an excessive overhang from the chuck, support the end with a three-point steady as near to the end as is practical.

The alternative to machining it is to simply file it square, something that sounds considerably more difficult than it is. To do so, lay the tube on the part-opened jaws of a vice, leaving just enough overhang for the amount that has to be trimmed off. Hold a scriber against the edge of the vice jaw, and carefully rotate the tube, ensuring that each movement is marked by the scriber in such a way that it lines up with the previous mark. After a complete rotation the scriber marks should meet, and the tube can be trimmed to them.

The next task is to drill a hole for a bush in the top of each tube; a little care is needed here to ensure the tube is not distorted by too much

pressure being applied, as the copper is thinner than that used for the previous boilers. In addition there is a danger of the drill snatching when it breaks through the copper. A small piece of emery cloth, or even ordinary cloth, can be placed between the drill and the work to prevent this from happening.

We now need to seal the ends of the tubes, and this can be done with the same copper as used for the barrels. Cut off a short length of the tubing, then cut it lengthways, and open it out to form a small sheet – though before doing so it should be heated until it glows a dull red colour and quickly quenched in water to anneal

While it is always best to flange the end plates of any boiler, in this case the tubing is of such small diameter that making flange plates creates its own problems. Some builders therefore may wish to just silver solder a piece of sheet copper across the end. The piece seen here that has been used for this purpose is from the tube; it was cut lengthways and opened out to obtain the required material. After cleaning in acid the plate will be shaped round the tube edge with a file in order to get a neat finish.

it. It is a matter of choice whether to make a former and flange the end plates, as was done with the demonstration model, or simply to lay the end plates flat and silver solder them to the tube ends; with tubing of this diameter the latter method is perfectly safe.

Before sealing the ends it is essential that the holes referred to in the previous paragraph are made: trying to seal both ends of a tube with no means for the hot air to escape, as would be the case if the holes were not made, is absolutely asking for trouble.

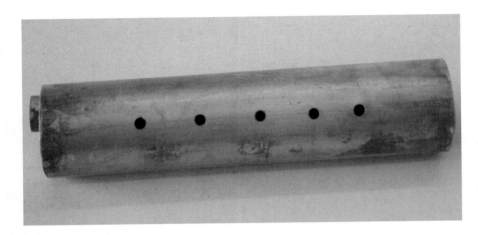

A series of holes are drilled lengthways along the barrel in preparation for the insertion of the water tubes.

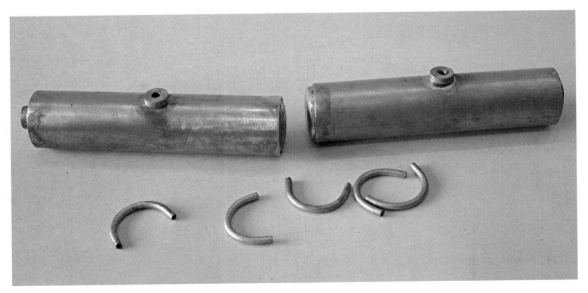

The water tubes are cut to size, annealed and then bent to shape; these ones have been pushed round a length of steel bar. It is possible to wind the water tubing round a former as one would when making a spring. The sections are then cut from that, but it can be difficult to get them an even length when doing it that way.

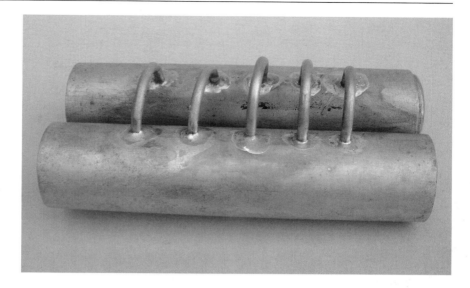

The underside of the finished boiler; note how the tubes are staggered in order to obtain maximum efficiency.

There are also five holes in each barrel to accept the cross tubes, and these must be evenly spaced, but the two sets should be slightly staggered to assist the water flow. The cross tubes should be carefully bent to shape, and it is best to anneal them and then shape them round a former of some sort – a short length of bar will do, but do take care to avoid kinks. They should enter the main tubes by no more than ¹⁄₁₆" (1.5mm), and should be silver soldered in place.

It should be possible to silver solder the whole boiler together with a DIY-type of blowlamp, but it will be necessary to ensure that the job is well packed with some sort of reflective material. It is usual to use reflective firebricks for this, but again we can improvise, rather than go to the expense and trouble of getting those. Make up a square of ordinary firebrick of the type used in domestic fireplaces, and in front of those insert a couple of pieces of stainless-steel plate; this can be obtained by buying a couple of cheap baking dishes, or something similar. The stainless steel will quickly discolour and will have to be re-polished after each heating of the job, as we are relying on it to reflect the heat back to the work.

The final task is to solder the bushes in place, and as before, it is best to tap them for about half their length when making them, and then use a piece of threaded steel rod to straighten them should they be out of line. Finally open them with a tap in order to get full-depth threads.

The plugs used for checking water levels have already been dealt with at length, and so need no repeating, except to say that on a boiler like this, the type with holes drilled down the centre is possibly better than plain ones.

A long, narrow burner is needed in order to get maximum efficiency from the engine, and there is no reason why this boiler, like all the others, cannot be used with a reversing controller, as in practice the operation is no different to that of a more conventional design. The design has certain advantages over the more conventional type as it is lower and also wider: this means it can be useful where headroom is limited, but width is not. In addition the design particularly lends itself to lengthening, as long as the burner is also lengthened, of course; unlike vertical boilers, where the heat loss limits the proportions of height to diameter, this one works at maximum efficiency throughout its whole length.

133

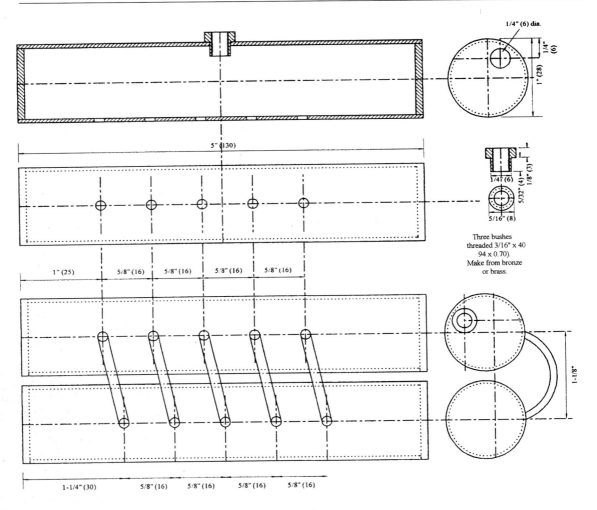

1/4" (6) dia.

1/4" (6)

1" (28)

5" (130)

1/4" (6)

1/8" (3)

5/32" (4)

5/16" (8)

Three bushes
threaded 3/16" x 40
94 x 0.70).
Make from bronze
or brass.

1" (25)   5/8" (16)   5/8" (16)   5/8" (16)   5/8" (16)

1-1/8"

1-1/4" (30)   5/8" (16)   5/8" (16)   5/8" (16)   5/8" (16)

Constructional drawings for the double-barrelled boiler.

19 Supports and Burners

Once a boiler is made it is of no use unless it can be correctly situated in relation to the engine, therefore a support that will hold it steady in one position is necessary: such a support can be easily fabricated from either thin sheet steel or aluminium. The type of boiler will dictate the support that will be needed – a horizontal one will require something to keep it at a suitable height to allow the burner to do its work – and to some extent will govern the measurements chosen: its size will also have to be taken into account; but in all other aspects the support will be of a generally accepted pattern.

HORIZONTAL BOILERS

A stand for a horizontal boiler consists of two sides and two ends, and the ends should be shaped so that the boiler can nestle easily on them, while giving support partially round the circumference as well as at the bottom. One end of the stand requires a section cut away to allow the burner to be placed underneath, otherwise both should be identical. Sides as a general rule are flat, and there must be a row of holes to allow air to be drawn in to give the necessary oxygen to the burner. For most purposes the more air holes there are the better, but if the boiler is being used in a model boat there might be a problem of excessive draught from the wind; it is not unknown for model makers to arrange some sort of sliding arrangement so that some of the excessive air is cut off. Very often the boiler will be open to admit air at one side only, keeping closed the

side from which the wind is coming. However, it is not the purpose of this book to delve into such complications; we are concerned only with the ability to support the boiler in such a manner that the burner can do its work efficiently.

Shaping the Ends

Shaping the ends of a stand for a horizontal boiler can be done either by hand using drills and files, or with a trepanning tool; in the latter case it is best to cut a complete circle from the sheet of metal and subsequently divide the metal in half centrally across the circle that has been cut out. Most boiler supports are made of very thin steel or aluminium sheet, but if a thicker material is used it is quite practical to put the metal on the lathe faceplate and use a boring tool to get the profile. Of course it is not a good idea to place the metal in direct contact with the faceplate, as this would damage it: the metal should be mounted on a piece of wood, which in turn is mounted on the faceplate, or if a big lathe is available it could be held in a four-jaw chuck. Medium-density fibreboard (MDF) or chipboard are ideal materials on which to fit the work; because of their smooth finish it is quite practical to use a good quality, double-sided adhesive tape to hold it in place. An old trick used by cabinet makers is also worth considering: they would stick a piece of newspaper to the work, and then stick the other side to the wood using ordinary woodworking glue; after machining, a quick soak in hot water was sufficient to part the pieces.

To shape the work by hand, the radius must

135

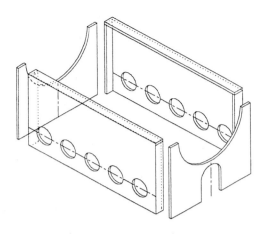

Drawing showing how a suitable stand for a horizontal boiler can be fabricated using sheet metal and small angled strip.

will just break into each other, and the part will fall out. The edge is then finished with a file.

To some extent the method used will depend on how it is intended to construct the support. It is quite practical to use a single length of material, and after shaping the ends, to bend it to form the completed support; the only thing left to do in this case is to solder or otherwise fix one edge joint. Bending metal accurately is not easy, however, and only very thin sheet should be used if attempting to make the stand in one piece. Stands can be made in a similar fashion, but instead of folding all four sides, a single side and end is made, with only a single bend to make in each section; this is an easier way, but of course while a bend has been saved, it now means that there are two joints to contend with.

It depends on individual ideas and wishes as to how the two ends and two sides are joined. Some will prefer to fold a section at the end of the metal at right angles, thereby forming a tag and either soldering or riveting that to the other section. How well this works out will depend on the thickness of the metal used; bending thin sheet is easy enough, and can

be marked out and then a coping or piercing saw can be used to cut it to shape. An alternative to this approach is to use a method known as chain drilling, which involves drilling a series of small holes around the area to be cut away; ideally they should be so spaced that they

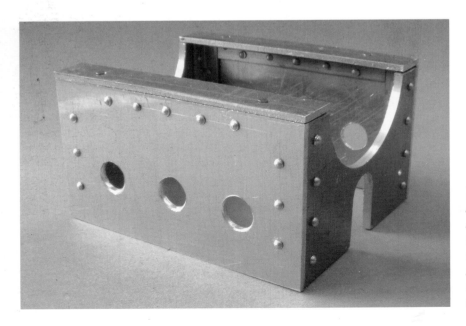

This is the stand depicted in the drawing and made by riveting four sides to pieces of brass angle. Brass snap-head rivets have been used for the assembly.

136

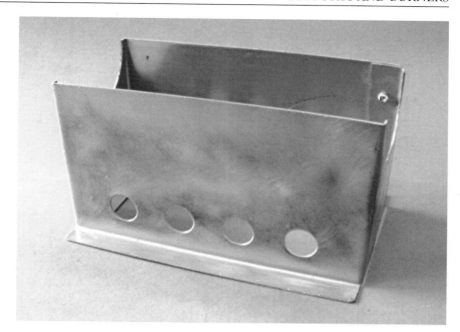

A stand for a horizontal boiler made from a single length of aluminium and pop riveted together.

This stand has been made from two separate pieces of steel sheet and joined with nuts and bolts.

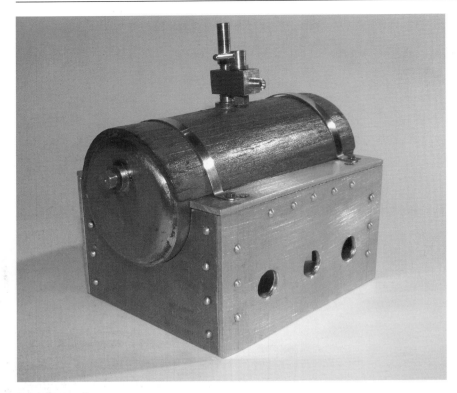

The stand made from four separate sections seen here with the Number One boiler fitted; it is held in place with two brass strips, screwed to the stand. The strips also secure the lagging, which has been made from mahogany strip wood.

A view of the other side of the stand to which the boiler has been fitted; note also the turret that contains the on/off regulator and the safety valve. In each photograph it can be seen how the long burner fits neatly into the set-up.

either be done with a bending press or by simply pushing the part to be folded, while the metal is held in the vice. If using aluminium to make the support, take care when it comes to bending, as some of the alloys will crack and break, rather than bend. Unfortunately aluminium alloys are not the easiest of materials to anneal in order to prevent this from happening, unless they can be placed in a heat-controlled oven. Any attempt to heat them with a blowlamp invariably results in localized distortion and buckling of the metal.

The type of rivet used to join the section is yet again a matter of personal preference: snap-head rivets will give the support a more professional-looking finish; pop rivets do not look so good on the finished product, but are very much easier to use. An alternative is to use screws, tapping one section for the screws to fit

into, and yet another would be to use small nuts and bolts to do the job.

If bending the metal to form tags is found to be difficult, consider putting pieces of angle at the corners and fixing the sides and ends to these; this method makes a good, strong job and saves the frustration of finding that the tag is not quite in the desired place and as a result the finished support stand is not square. The use of small angle section will almost certainly guarantee the end result to be square and accurate, and there is not a great deal more work involved in doing it that way.

VERTICAL BOILERS

In general it is easier to make a suitable stand to support a vertical boiler than it is for a horizontal one. All vertical boilers should be

RIGHT: A simple stand for boiler Number Four; the same format will also suit the Number Two boiler. At this stage the boiler has not been lagged, but note the fact that the regulator has been constructed to work in a vertical position, by simply moving the connection. A round burner is fitted below the boiler.

FAR RIGHT: A second view of the vertical boiler stand, made from a food can; it is held to the boiler by three bolts passed through the section of the barrel below the end plate, and secured with nuts.

constructed in a manner that allows an overlap of the barrel below the bottom plate; this will give room for suitable fixing points for the support. The actual support can take the form of a wrap-around section, or of a tripod; in the latter case the legs can be fitted to the overlap of the boiler shell with nuts and bolts. All that is required in that case is three pieces of strip metal, and once they have been fitted, the job is complete.

The alternative is to make a wrap-around section and screw that to the overlap. When dealing with a vertical boiler it is just as essential to allow sufficient ventilation to keep the burner going as it is with a horizontal one, and suitable arrangements must be made for this, by making suitable airways in the base. Old food cans are particularly suitable for making this type of base, as the metal is already curved.

The photographs and drawings demonstrate quite clearly what is needed, and everyone will have their own idea as to finish.

BURNERS

It does not take too much reasoning to realize that a long burner will be needed for a horizontal boiler, and a round one is preferable for the vertical type. The long burner shown in the drawings is suitable for both horizontal boilers; it should be made of brass, and the outlets fitted with wicks. Suitable material for a wick is available from good ironmongers or some model shops, particularly those specializing in model boats. It is essential that the holding tank is not higher than the top of the outlets, otherwise the fuel will run out. It does expand with the heat once the burner is alight, and it is very dangerous if lighted fuel does spill out.

Although only a single round burner is shown for the vertical boilers, there are in fact three different types, as there is the lozenge tin used in Chapter 4, and another round one used for the locomotive in Chapter 20. Both of these are quite capable of running the vertical boilers.

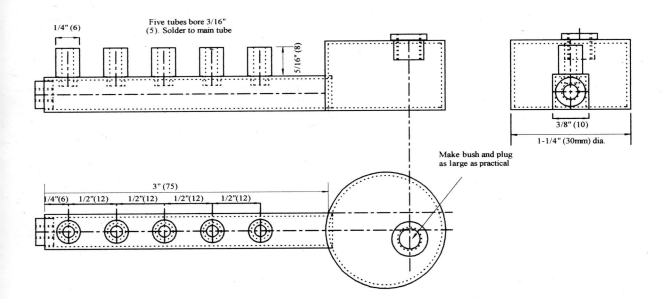

Burner suitable for all horizontal engines.

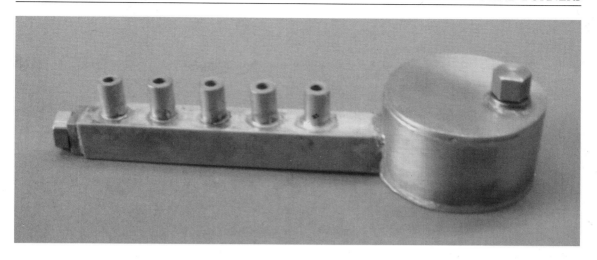

The finished long burner, a fairly simple soldering job.

The lozenge tin has the advantage of simplicity, with just one hole to be drilled; also because the lid of the tin is removed for filling it is very easy to fill. All the other burners really need the use of a small syringe in order to fill them without the fuel spilling.

The burner for the locomotive is different inasmuch as there are no burner tubes, the combustion taking place at tank level via a number of very small holes. Strips of cotton act as wicks, but most of the heat comes from evaporation of the fuel that can travel past the wick and burn as a gas rather than a liquid. This results in considerably more heat than the more usual arrangement, and also a burner that has much less height than usual. It is the lack of

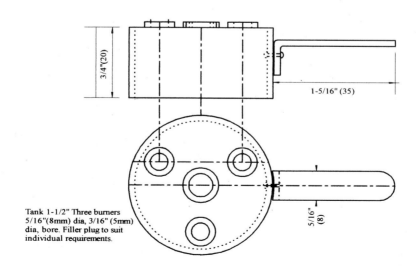

Tank 1-1/2" Three burners 5/16"(8mm) dia, 3/16" (5mm) dia, bore. Filler plug to suit individual requirements.

3/4"(20)

1-5/16" (35)

5/16" (8)

Burner for vertical boiler.

A round burner suitable for use with either of the vertical boilers.

height that makes it useful in the locomotive, but at the same time because of its efficiency it is quite suitable for the ordinary boiler.

Fuel and Combustion

The fuel used is methylated spirits, a colourless liquid that has a dye added to it; the reason for the dye originally was to try and stop it being drunk by people who became addicted to it. It is doubtful if there are many such people around these days, but the dye is useful as it allows it to be seen – but only just – when it spills. When alight, in bright sunlight the flame is hardly visible, and this can be a problem as it is all too easy to burn one's hands because it is believed that the burner is not alight. The flames fortunately do not give very serious burning, but it is not something one wants to happen. It is therefore suggested that if the engines are to be operated by children, fuel tablets, sold in model shops, should be used as there can be no spillage with their use. They do not give the same amount of heat as the spirit, and leave a heavy soot deposit on the underside of the boiler that needs to be removed from time to time as it impairs the efficiency of the

boiler. A small metal tray should be folded up, and two or three tablets placed on it for maximum safety.

Whether spirit or tablets are being used, the flame should be set so that the top of it just touches the base of the boiler and spreads a tiny fraction. If it is set too high the combustion will not be complete, and the boiler will not work efficiently. If set too low the flame will be at full heat but may not be sufficient to reach the boiler – although as a general rule, too low is better than too high.

Finally a word about the fuel itself

A great deal of methylated spirits now sold does not generate a great deal of heat; it is not sold with the intention that it should, and the manufacturers have therefore adapted it accordingly. This particularly applies when it is bought at a chemist, and therefore it is advisable to search around to get some industrial quality spirit. Good engineering suppliers stock it, and so do many model shops, so if some difficulty is experienced in raising steam, first suspect the fuel quality, and try a different supplier.

PART FOUR
FINALE

20 Vertical Boilered Locomotive

The theme throughout this book has been the construction of very simple models that are easy and inexpensive to make, while the finished product is a working artefact and hopefully attractive to look at. This final chapter deals with something entirely different, because although the basic theme of oscillating engines is continued, it is not quite such an easy project. It is to be hoped that readers who have made all, or even most of the models described so far, and who did not have the expertise to do so before, will now be able to complete this one. It will also supply an interest and a challenge to the more experienced operator, and the finished model is quite delightful. The locomotive as drawn is designed to run on '0'-gauge track, and because railways in that gauge are invariably measured in metric form, there is a change in format on the drawings, with the metric measurement shown first followed by an imperial figure in brackets.

Readers should be warned that construction of the model calls for the use of very small bolts and screws, and therefore involves tapping sizes as small as 1.6mm, or 10BA. Providing care is taken when making the threads using these small taps and dies, it is not difficult. But the watchword has to be 'CARE', otherwise taps will break in the work and the part being worked on irretrievably damaged. It is surprisingly easy to set a tap in the work at an angle, and that is possibly one of the biggest causes of broken taps.

The engine is based on what was once a common type of locomotive in Britain: the vertical-boilered design of the De Winton Company. Most of the locomotives of this type were designed for narrow-gauge operation, and there were many built. A number have been preserved and are to be seen at various preservation sites and museums throughout Britain. Some of these are still in use, though nowadays tend to pull passengers rather than the various freights that they were originally designed to deal with.

THE FRAMES

The frames are made from mild steel sheet or strip, and have been left in as basic a form as possible while having sufficient embellishment to make them look right. It will be seen that

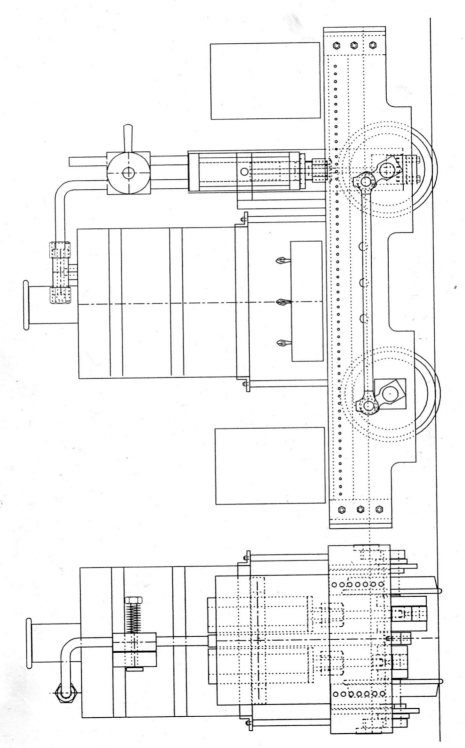

General arrangement of vertical boilered locomotive.

144

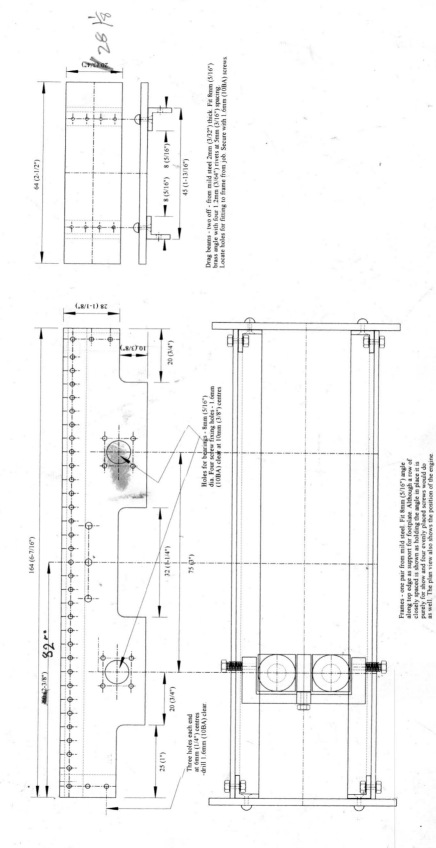

Frames for vertical boilered locomotive.

View of the engine, in which the tripod arrangement for raising the boiler to get clearance for the burner can be clearly seen.

The opposite side of the engine, showing the arrangement of the reversing controller and piping from the boiler. Note that an additional steam outlet has been fitted, but is not in use.

along the top of the frame is a row of closely spaced rivets: these are holding strips of brass angle, which in turn supports the boiler and footplate. The number of rivets may appear to be excessive, but they are there because it was a feature of the original locomotives. It is certainly not necessary to use that number, and the angle can be supported with three bolts without detriment to the strength of the model.

The best way is to make them as a pair, and to do this, start by drilling one of each of the holes for the buffer beams at each end of the chassis. Use the holes to bolt the pair together, and then complete the other holes and cutting out. Repeat the operation on the buffer beams, and then assemble the necessary pieces of brass angle that hold the frames together. Note that no provision has been made for couplings or buffers, this being left to the individual builder. Those with model railways on which the model can run, will have a standard type of coupling and will wish to use that on this model. Likewise with buffers: most engines of this type had what were known as 'dumb buffers', which were nothing more than massive pieces of hardwood. There were, however, examples with sprung buffers, and even some that had none at all.

The tops of the frames are covered, and after assembly a sheet of thin brass can be used for this purpose. A gap has to be left for the engine to pass through, and another for the burner; readers may therefore find it easier to use two separate sheets, and this is how the engine in the photographs was built.

AXLE BOXES

Because the frames are of thin section mild steel, it will be necessary to fit axle boxes, and these should be made of bronze if possible, the alternative being brass. The drawings show them as square, because that is how they look on the full-size engine. They are not really axle boxes at all, just bearings, and so if anyone wants to make simple round bearings and use them, that is fine: the engine will run just as well, even though they will not look quite as good as square ones.

AXLES

Two different axles are required: one is a straightforward case of machining a step at each end of a length of round bar; the other is far more difficult, and the success of the running

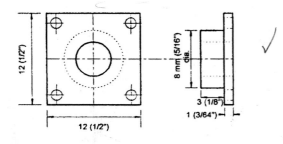

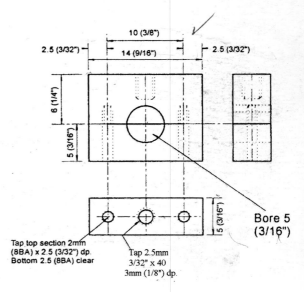

ABOVE: Wheel bearings: four off from brass or bronze – bore 5mm (³⁄₁₆"). Holes for bolts at 10mm (³⁄₈") ctrs. 1.6 (10BA) clear. Alternatively thread 8mm (⁵⁄₁₆") section and secure with thin nuts.

RIGHT: Big end bearing: two off from bronze or brass. Note split along centre line.

ability of the locomotive will depend on how well it is made. It is, of course, the crank axle, and whereas until now crankshafts have been very simple affairs, this is not possible in this case. However, with care, making a crank axle is not that difficult; full measurements for the bits are shown, and there is another drawing that shows how to assemble the axle.

Start by machining an identical axle to the non-cranked one, and then drill the four webs; the drilling must be accurate, otherwise the thing will not work. Push the completed axle through four of the holes, carefully checking the spacing as everything will depend on getting that right. Before this part of the

assembly, run a little retaining compound round each hole so that when the axle is pushed through and checked for spacing, the adhesive will set and hold things in position. Before it does set, though, put a length of steel rod through each of the other two pairs, again using a retaining compound to eventually hold things in position. Give the assembly plenty of time to set, at least twenty-four hours, that is, and longer if possible. Finally, when absolutely certain nothing is going to give way, put a hacksaw through the main axle where it passes between the webs, and clean the inside of them with a file.

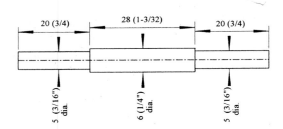

Axles: two off – mild steel.

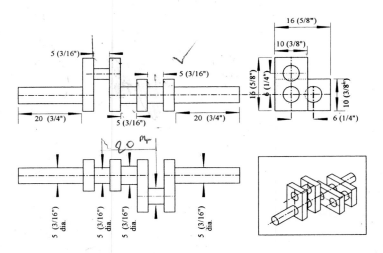

Crankshaft: from mild steel. Fabricate or machine from solid. Inset shows set up for assembling the crankshaft.

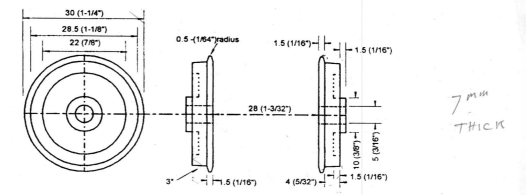

Four wheels: from mild steel.

WHEELS

The wheels are machined from mild steel bar and involve nothing more than drilling, shaping and parting off; the latter may cause some people palpitations, as the parting tool must be run to considerable depth. Work of this type calls for a sharp parting tool and plenty of cutting fluid; the movement through the metal should in theory be one of steady progress, keeping the same pressure on the tool throughout. Having said that, and without a doubt it is the correct way, the author knows of one modeller who does exactly the opposite: he advances the tool a few thousands of an inch, or hundredths of a millimetre if you are a metric person, and then withdraws it, and he keeps repeating this a-little-at-a-time movement until the job is done, and it seems to work quite well.

Yet another method is to take the parting tool in a short distance, withdraw it, and then take a second cut of nearly but not quite the same depth, about half the width of the parting tool away. Then take the tool back to its original position and make another cut, then repeat the operation. The idea is to run two cuts, and in doing so to reduce the build-up of heat, which is the main cause of the tool snatching, the heat causing the metal to expand and clamp on to the tool.

Finally for the really faint-hearted, the shaped section can be sawn off, and when all four have been completed, they are replaced in the chuck in reverse position and faced off. Even if the wheels have been parted off, facing off the back is not a bad idea as it is possible to make sure that the back of each wheel is an identical width, something that cannot always be done with complete accuracy when parting off.

THE ENGINE

This is a double-acting engine that differs, but not substantially so, in the method of manufacture to those already described. What are the differences? First of all the port block is also a bearing that runs on the centre journal of the crankshaft. Because we need to be able to get the shaft in place, the hole acting as a bearing has to be split across and the two sections joined with small screws. To do this, start by making the holes for the screws and drill them, at this stage just tapping size only. Make a cut where the bearing is to be, separate the two sections, tap the larger one, and open the lower holes to clearance size. Screw the two parts together, and drill a hole exactly on the cut line; it goes without saying that the cut line must be made very carefully as any deviation will make it difficult to insert the shaft.

The ports pass right through the block and

149

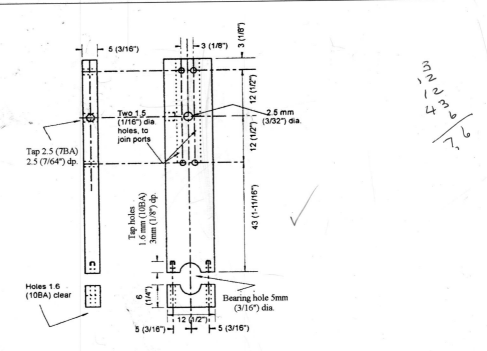

Port block or body: one off from brass. Ports 1.5mm (¹⁄₁₆"). Join parts with 1.6mm (10BA) bolts.

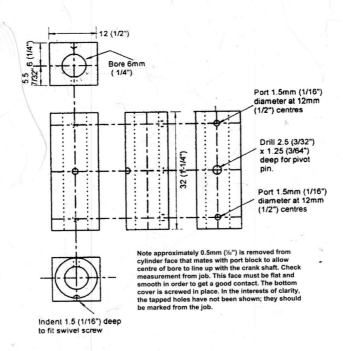

Cylinder blocks: make two from brass or bronze. Polish port face.

are joined with holes drilled down from the top, which again is something we have done before. The pivot hole also passes right through the block, as the cylinders are going to swing on either side of the block. Apart from rubbing it on a sheet of fine emery cloth, using a figure-of-eight motion in order to ensure that both sides are perfectly flat, that is all there is to it.

THE CYLINDERS

The cylinders in this case start life as square bar, which means there is no port face to be soldered on. The first thing is to bore them centrally and to drill the ports and pivot hole, and at this stage that is all that is needed.

PISTON ASSEMBLY

We now come to another major alteration: the piston and piston rod follow the same pattern as with other engines, but the big end bearings need to be split in the same way as the port block. First of all, thread them to accept the piston rod, and then follow the sequence of operations given for the port block. Again we have a situation where accuracy is very important, particularly when it comes to making the saw-cut. This is a rather awkward job, because the pieces are so tiny. But should anyone making the engine have one of the little hand-held mini drill/grinders, it is possible to buy small circular saws to fit them, and they can be a great aid when making the cut as it is easier to make a straight and even cut with them.

Pistons: 6mm (¼") dia. Tap 7BA (2.5mm).
Piston rods 2.5mm (³⁄₃₂") dia. 63mm (2-½")
long. Check length from job.

CYLINDER COVERS

Some people may be wondering why the cylinder covers have not yet been mentioned, and this is because the piston and piston rod may help to locate them. We are in the same situation as with the other double-acting engine, 'Douglas' in Chapter 9: the bottom covers have an extra threaded piece to take the gland nut, but whereas that was soldered in place, it is much wiser in this case to screw it, so that it can be removed if problems occur at a later stage. Anyway, make the covers as before, and by all means solder the top ones in place, if that is preferable – or perhaps you may wish to bolt those on as well. However, if they are to be soldered, do not fit them until the engine has been assembled to the frame at least as a trial, because it is necessary to be able to see inside and watch the piston movement during the assembly, and this cannot be done with the top covers on.

It is now necessary to drill the bolt holes in the covers, but again there is a little problem. If the cylinders are left as they are, the big ends will not line up with the bearings, so a little has to be removed from one edge to get matters right – and while on the drawings the figure is shown as 1mm or ½", this is only nominal, and it will be necessary to work it from the job.

Put some form of marking fluid on the two crankshaft journals, and make a mark at the exact centre of each. Assemble the shaft to the port block, and clamp the cylinders to it, as if the engine was being assembled; it is preferable if this can be done with the pivot pin in place. Take a length of steel rod of the same diameter as the piston rod, but a little longer. Thread the end so that it will fit in the piston and then, using the lathe, make a sharp point on the other end. Put it on a piston and push the piston into the bore until the point of the rod touches the journal. Rotate the crankshaft, and the point will leave a mark round it that can be measured from the central one that had previously been made. That is the amount by which the port face needs to be reduced, a task that is best carried out in the lathe using the four-jaw chuck, or in a milling machine if one is available. If neither is available it will be a case of careful filing and rubbing on sheets of abrasive paper; in the latter case, file away a tiny section either side of the pivot, about 1mm or ½₂" deep. This recess is an aid to the alignment of the cylinder if the face is not quite accurate,

and can also help prevent some of the rocking motion likely to occur when rubbing the cylinder on the abrasive cloth.

The end result must be the point lining up centrally on the mark on the crankshaft journal. Removing a little from the side of the bearing can rectify any tiny errors, but do not, under any circumstances, thin it too much.

We can now get back to the cylinder cover, and should start by making the holes for the bolts; incidentally, in full-size practice bolts would never be used, it would be studs and nuts, so some constructors may wish to follow that practice. Transfer the holes to the cylinders, tap them, and bolt the covers in place; finally, remove the edge that is overlapping until it is flush with the edge.

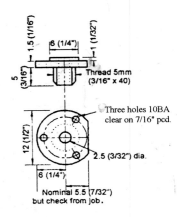

Cylinder bottom covers: two off – brass or bronze.

THE YOKE

The most unusual thing about the engine is the method of keeping the cylinders flush to the port block, when room is so limited; this has been solved by using a yoke, and passing screws through that to adjust the piston friction. Although it all sounds very unusual, in fact it is quite an easy arrangement to make, and the drawings show all the necessary details. In turn the yoke is either screwed or soldered to a bracket on the frame. It is necessary to adjust the position in which the yoke is finally fixed so that the pistons reach the correct height.

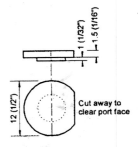

Top caps: make two from brass. Soft solder to cylinder block.

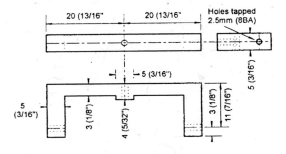

Yoke: one off – brass. Can be soft soldered to port block or use hole shown to secure with 2.5mm (7BA) screw.

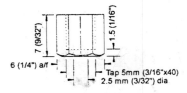

Gland nut: two off – brass or bronze. Fit packing or "O" ring.

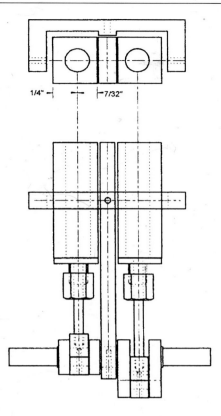

1/4"   7/32"

ABOVE: Layout of cylinder assembly and yoke.

RIGHT: The workings of the boiler.

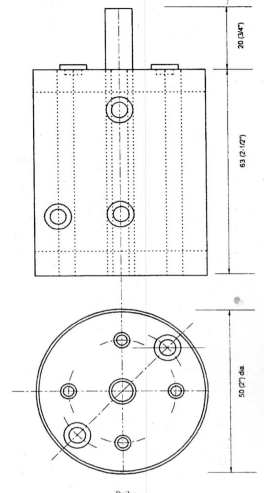

20 (3/4")

63 (2-1/2")

50 (2") dia.

Boiler.
Make from 1mm (20 gauge) copper - silver solder
all joints. Four tubes 5mm (3/16") dia. at 30mm
(1-1/4") pcd. Centre flue 8mm (5/16") dia. - fit five
bronze bushes as shown. Support on frame with three
1" (25mm) x 3/32" (2.5mm) pillars.

THE BOILER

The boiler is a truncated version of the Number Three boiler, and has been shortened in order to give the engine a similar appearance to the engines generally used in Britain. Without doubt a taller boiler would give a longer running period, and this type of engine in the United States invariably had taller boilers than the British ones, so the use of Number Three boiler would not be wrong.

The boiler on the model shown has been lagged with an old food can after wrapping thin fibre-glass sheet round it, but wooden lagging would not be wrong; both are held in place with brass bands, the lower of which has three

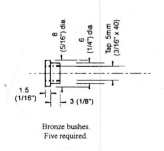

8 (5/16") dia.

6 (1/4") dia.

Tap 5mm (3/16" x 40)

1.5 (1/16")

3 (1/8")

Bronze bushes.
Five required.

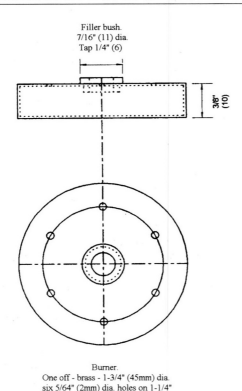

Filler bush.
7/16" (11) dia.
Tap 1/4" (6)

3/8" (10)

Burner.
One off - brass - 1-3/4" (45mm) dia.
six 5/64" (2mm) dia. holes on 1-1/4"
(32mm) dia. pcd.

Drawing showing construction of burner.

The rather unusual burner with six small flame outlets. This has the effect of the fuel vaporizing on the surface, and giving greater heat than if just burnt with a wick. The top surface is marked, showing where this type of construction has resulted in this happening.

brackets that are bolted to short pillars that lift the boiler clear of the burner, and in turn are screwed to the frame.

SUPERSTRUCTURE

These engines didn't really have any superstructure; they had a small water tank at one end and a coal bunker at the other, and the driver, who also did the stoking, stood in the open space between. In our case the two fitments can be included for appearance, but they will not have any real purpose. If they are wanted, they can be fabricated from old food cans or brass sheet; even blocks of wood could be used.

There is a reversing valve that also acts as a regulator and stands on the footplate; it could have been placed on one of the end fitments, and some builders might find that arrangement more convenient. The regulator is exactly as described previously, and the necessary connections with the boiler and engine can either be soldered in place or fitted with union nuts.

THE BURNER

Only a small burner is used, but it is sufficient to give a running time of around a quarter of an hour or so. It can be fabricated from thin brass sheet, which can safely be soft soldered together. To make it, take a strip of the brass and join it end to end to form a circle. Solder the top section on; there is no need to cut it perfectly round before doing so, as it is better

154

trimmed to shape after soldering, when it is far easier to do.

The next job is to drill for the bush that accepts the filling cap, and the holes where the actual burning takes place. There are six holes that act as burners, and they should be drilled on a concentric circle; they are 2mm (5/64") in diameter, which may sound rather small, but in fact the burner works very well indeed and might be worth considering for the other vertical boilers.

The bottom of the burner can be soldered in place when the holes have been completed; on the bottom is a small piece of brass strip. This is carefully cut to fit in the gap in the footplate so that when the engine is running, the burner will not wander all over the place. The burner slides under the boiler, which is supported on three small pillars, to give it sufficient room. In turn, three small brackets are soldered to what amounts to a boiler band, the brackets fit the tops of the pillars, and the assembly is held in place with 10BA or similar screws. A certain amount of adjustment of the boiler height is available by altering its position within the band, and this will allow things to work at their maximum efficiency.

Steam is taken from the top of the boiler, via the reversing valve to exhaust; the position of the exhaust pipe will be a matter of individual choice, and unfortunately the construction of a boiler in conjunction with an oscillating engine invariably results in some sort of compromise when it comes to disposal of the exhaust.

CRANKS AND COUPLING RODS

As the engine is outside framed with the wheels on the inside, the usual practice of casting a web to which the coupling-rod bearing is attached is not possible. Therefore separate cranks have been used, and these should first of all be bolted in position and then a hole drilled between the crank and the axle for the insertion of a pin. Coupling rods on these engines were never fluted, and frequently consisted of little more than a heavy bar of steel. A major difference on the model is the fact that the rods are fitted via bronze pins, instead of the more usual crank pin.

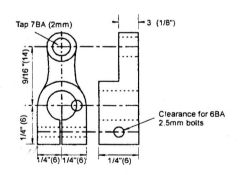
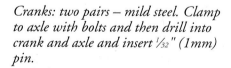

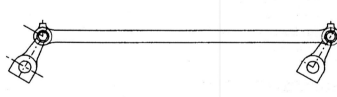

Cranks: two pairs – mild steel. Clamp to axle with bolts and then drill into crank and axle and insert 1/32" (1mm) pin.

Coupling rod arrangement: make from steel to fit.

Glossary

Below is a short glossary of abbreviations and terms that might not be familiar to readers without an engineering or technical background.

Term or abbreviation	Meaning
a/f	across flats
af	across flats
big ends	bearings that fit on crank pins
c/bore	counter bore
csk	countersunk
dia.	diameter
dp	deep
ht	height
lnth	length
m/s	mild steel
pcd	pitch circle diameter
port block	frame against which the port face runs
port face	flat section on cylinder with steam ports
psi	pounds per square inch
r	radius
retaining compound	a high strength anaerobic adhesive
swg	standard wire gauge
tpi	threads per inch
SWG	Standard Wire Gauge

Index